# FACING THE CHANGE

## PERSONAL ENCOUNTERS WITH GLOBAL WARMING

EDITED BY
STEVEN PAVLOS HOLMES

TORREY HOUSE PRESS, LLC

SALT LAKE CITY • TORREY

First Torrey House Press Edition, October 2013

Published by Torrey House Press, LLC
P.O. Box 750196
Torrey, Utah 84775 U.S.A.
www.torreyhouse.com

International Standard Book Number: 978-1-937226-27-5
Book and cover design by Jeff Fuller, shelfish.weebly.com


The following selections appeared previously in the indicated venues:

"About the Weather" by Harry Smith, in Harry Smith, *Little Things,* Presa Press, 2009. Copyright © 2009 Harry Smith

"Snowshoe Hare" by Roxana Robinson, as "A Hare Out of Season" in *Boston Globe*, February 16, 2004. Copyright © 2004 Roxana Robinson

"On the Eve of the Invasion of Iraq" by Todd Davis, as "Upon the Eve of the Iraqi Invasion, My Wife Says" in Todd Davis, *The Least of These,* Michigan State University Press, 2010. Copyright © 2010 Michigan State University Press

"Be Prepared to Evacuate" by Tara L. Masih, in *Amoskeag: The Journal of Southern New Hampshire University,* 2009. Copyright © 2009 Tara L. Masih

"The Things We Say When We Say Goodbye" by Alan Davis, in *The Sun* no. 393, September 2008. Copyright © 2008 Alan Davis

"To Wit, to Woo" by Kathryn Miles, online at *Terrain.org* no. 24, fall/winter 2009. Copyright © 2009 Kathryn Miles

# Table of Contents

## Part III: Revolutions

Sara L. Marsh
Boston Nature Center
Dec. 8, 2013

Steve Holmes

# Introduction

Unlike most things that you've probably read about climate change, this book is filled not with science and politics but with stories and poetry. The writings collected here are not intended to convey new facts, to prove any particular theory, or to make predictions, though you may find any or all of these in them. Rather, they convey something more amorphous, individual, and emotional, even spiritual: what it *feels* like—for these authors, at least—to be living in a world in which the climate seems to be changing as a result of human action, and what those changes and those feelings *mean* for us, for our understanding of our place in the world, for our perceptions of ourselves as ethical beings, for our hopes and fears for the future. And though we think you'll learn a lot about the world from these stories, what's most important isn't whether you believe or agree with these reports from other fields, but whether they resonate with what you yourself are seeing and feeling on your own turf, in your own life.

Since the following essays, poems, and short stories were written at different times over the past ten years, in different places and in response to different experiences and events, you may or may not see the same things as are described here when you look out your own window. Moreover, we all should keep in mind the distinction between *weather*—i.e., what is happening today, or this season or year—and *climate*—the longer-term patterns unfolding over decades. Indeed, as I write these words in the winter of 2013, my fellow Bostonians and I are slogging through yet another cold, wet, snowy winter, no different from many in the past, with surely many more to come; but I still find myself feeling a sense of surprise, and relief, that such a winter can follow on last year's unseasonable warmth and the wimpy winters of the past few years. At the same time, reports from elsewhere—of floods, drought, tornadoes—remind me that even if things seem normal where I am right now, strange things are still happening in the world at large. And although anomalous weather events can't be tied scientifically to global warming, they often are *experienced* as part of the same emotional reality, evoking and contributing to a deepening sense of unease. Amid such uncertainty and unsettledness, each of us is challenged to put together our own experiences with what we hear from others into a larger picture of what's going on, and to think and feel our way beneath the facts to a deeper sense of what it all might mean.

In recent years, in the face of emergencies from fire to terrorism, many of us have learned to value the first responders, those equipped and resolute people who run toward rather than away from a disaster, who brave the fire and smoke and confusion to do what they can to help people in need. The writers in this collection are our emotional and cultural first responders to climate change—the ones who, with skill and insight, are showing up at this disaster, still in the making; who brave the fear and guilt and confusion to do what they can for people in need. And we are all in need.

On a personal level, over the years of working on this project, I can't count the number of times when, in the midst of the activities of an ordinary day—walking outside to a soggy snowfall, perhaps, or to a hot morning, doing the laundry or mowing the lawn—the words and images of these authors have come back to me like well-known songs, guiding me to broader understanding and deeper insight; to tears, to anger, to despair; sometimes, indeed, to a more honest and grounded depression, but at other times to the realistic courage and renewed commitment that come from a sense of companionship and community. That's the kind of book this is—one to be not just read but lived with, one whose heart unfolds further with time and experience.

So, as you read this book, we invite you to put aside for a moment any scientific debates and political arguments you may have heard about global warming, and open yourself to human stories and to poetic imagination. If you wish, imagine yourself somewhere you feel comfortable and relaxed, awake, and alive—maybe curled up in a favorite chair, maybe out for a walk in a favorite field or forest—talking with a new friend, someone with whom you've enjoyed a few casual conversations in the past and with whom you're glad to have the chance to talk at more length. You find you have a few things in common, some shared joys and struggles around family and work, maybe a common interest in the natural world, a love of words and sly humor. As your conversation grows deeper—as you settle deeper into your chair, or venture farther into the forest—there comes a pause, not because you don't have any more to talk about, but because you trust each other enough to stop talking for a while, to simply be together in these familiar and beloved surroundings. In the silence, you notice your own breath, feel the warmth in your legs and arms, feel the beating of your own heart. Then, out of that warm, companionable quiet, one of you starts talking—you may not realize who—about something else you have in common, something you both have thought about but don't quite have the words for, something you've often wanted to bring up in conversation but for which you never felt the time was right. Looking in your new friend's interested and caring eyes, you think:

*Well, now's the time; it's now, or never.*

Together, eager and afraid, you plunge into your stories.

# Part I
# Observations

*What does global warming look like, up close? Do you know it when you see it? How can you have a personal experience of a planetary phenomenon? While it's easy to notice the obvious things—bigger and more destructive storms and floods, drought, flowers blooming too early or migrating birds staying too late—it's harder to fit particular instances into a larger picture of what's happening in the world, much less to come to some understanding of what it means, intellectually or emotionally. In Part I, our authors grapple with these and other questions—without always finding answers—as they share their own experiences of unusual weather patterns and their effects upon animals, plants, and humans, primarily in the United States. Chapter One, "Strangely Warmed," begins with the subtle, nagging unease and denial that many of us have felt on an otherwise unremarkable winter day, and then moves through the year to increasingly dramatic and powerful events such as storms and coastal flooding. Chapter Two, "Species Out of Joint," offers examples of how longer-term changes in weather patterns are affecting particular species of animals and trees, exploring what happens when species are out of place as a result of—or even in an attempt to combat—global warming. Chapter Three, "Bearing Witness," focuses on one species whose fate serves for many as a sort of icon of the world to come: the polar bear.*

## CHAPTER ONE

# Strangely Warmed

## About the Weather

Harry Smith, *Maine*

Oh sure, I have seen winter thaws before,
and once or twice a season would seem right.
Last year was strangely warm, bereft of snow;
so far, this winter has more thaws than cold.
Ice fishing opened with no ice.
This morn the promised snow has changed to rain;
steam rises as the thin snow cover shrinks,
exposing tussocks of the meadow voles.

When snow returns late in the windy day,
I feel relieved, elated, yet upset—
weird weather all over the world. I hope
another spell of normalcy sets in.
Even five inches of ice on the lakes
would let us pretend that everything is right.

## Snowshoe Hare

Roxana Robinson, *Maine*

January, and twenty-seven degrees. Down in the cove the water is emerald green, as though it were midsummer, but we know it's not. In the garden, the shallow puddles from the last rain are frozen solid. We've had some mild weather lately, but now it's over: the air tastes like iron, and the sky is low and grey. Snow is imminent. We're all waiting for it.

The garden is ready, everything in it has been cut down clean. A single clump of tall grasses still stands, its pale dry stalks whispering and rustling in the wind. Everything else is gone, down to the ground. The short clipped stems are covered with pine boughs. The garden is bare, waiting for snow.

Also waiting for it is the snowshoe hare who's sitting inside the walled garden, against the concrete base of the house. During the summer, he might easily be invisible there, at the back of the deep beds, shielded by the heavy foliage, the big curved hosta leaves, the dense clumps of ferns. He might be invisible there right now, if it were a little earlier, or later. At this moment, though, he couldn't be more glaringly visible, because he's made the big mistake not only of taking shelter against a bare wall, in a flat empty garden, but of disguising himself as a snow bank, in a brown landscape.

The snowshoe hare (*Lepus americanus*) has a system that's exquisitely attuned to lowering temperatures, diminishing light, and the arrival of winter. Sometime in the late fall, this hare's summer-brown coat very sensibly turned white. Hares don't dig burrows, and in the winter, a hare may shelter beneath a log, or in the tunnels made by heavy brush covered with snow, though often he will snug himself up against something solid: a tree, a bank, a log, a weather break. Exposed in the open like that, a brown coat against a white field would be a dead giveaway. So, somewhere along the evolutionary timeline of *Lepus americanus*, an ingenious color change system developed, for survival—the shaft of each hair changes, seasonally, from brown to white, and back again.

Our wall seems to be a safe place for this one. It's sheltered by the surrounding fences, and it's quiet. The hare sits completely still, against the concrete, as calm and secure as though his bold white coat blended perfectly with its scumbly brown surface. It's not his fault we have no snow. He's near the front door, and we go in and out, peering discreetly at him.

Our hare sits bunched in a neat rounded dome, the curve of his back high. I had expected a hare to be long and low, but he is short and plump, like a teapot. His dark rounded eyes are brown, with a faint tinge of red. His front paws, just visible beneath his snowy chest, are oddly brown, as though he's got on someone else's shoes. His narrow ears are velvet brown, and there is the sheen of the shadowy brown undercoat, beneath his snowy overcoat. His nose, too, is a blotch of neutral brown, and there are a few vague streaks of brown on his shoulders: the broken lines of camouflage. He's watching, his dark eyes on me, but he is completely and utterly still, despite our comings and goings. His ears are tilted slightly up—he's listening—but they don't swivel: movement would give him away. Immobility is his protection, it's movement that draws the eye. His stillness says simply that he's not here; there's nothing here.

His color-coding, too, is part of the hare's survival strategy. This weather, though, is dangerous for him: he can't adapt to unseasonal fluctuations, he can't change color every time the temperature rises in January. The snowshoe hare, like most living organisms, has evolved over time in relation to a reliable weather cycle. Living things rely on a dependable climate, whether it's temperate, desert, or arctic. Deciduous trees, for example, cannot tolerate an early, unseasonal snowfall, while they still have their leaves. The weight of the snow on the leaves is too heavy for their branches to sustain, and they'll break. The loss of large branches will compromise the tree's health. Fruit trees, which flower in the spring, require calm, mild spring weather during their blossoming season. Bees won't fly in high winds, which will rip the blossoms off the trees, so the chance for pollination—and a fruit harvest—will be lost. Warm weather in the middle of winter acts as a signal for plants to use their stored energy and send it upward, producing tender green shoots. A subsequent freeze will kill the shoots, and the plant will need to muster the energy to start over again, later. The energy spent on the winter shoot was wasted. The second or third try may produce a plant that's small and stunted. Energy is the elixir of life in the natural world; no organism can afford to waste it. Successful survival strategies don't include false starts.

Unseasonal storms and temperatures put stress on agriculture and fruticulture and silviculture, as well as on wildlife, and plants and animals under stress have a higher risk for disease and infestation. Historically, there have always been unseasonal weather events, but historically they were unusual—that's why they were recorded. They weren't the norm.

The emission of greenhouse gases disrupts the layer of atmosphere in which weather patterns are formed. Increasingly, we're evolving a pattern of unreliable weather, one of unseasonal fluctuations, extreme storms, droughts, and floods that threaten the vigor, reproduction, and survival of plants and animals.

Our snowshoe hare waits, motionless, against the wall. If it snows again, as it normally would in January, he'll have protective cover, and the chances will diminish of his being seen by the coyote who trots past, down by the cove, along the shoreline, in the late afternoons. But if we have another winter of fluctuating weather, alternating between hard freezing and unseasonal warmth, the snowshoe hare—and all the trees on the property—and on the island—and in the state—will be in trouble.

## On the Eve of the Invasion of Iraq

Todd Davis, *Pennsylvania*

My wife says
there's a sadness
to this place, to the red
fog of maple buds,
their slow scale
up the sides
of these ridges.

The trees are too young
and certain slopes stay open
after they're clear-cut,
nothing but garlic mustard
and Japanese stilt-grass.

The air's too warm,
invasive as well,
and winter is wrapped
in less and less white
each year.

My sons miss
sledding, but don't
mind playing
in shirtsleeves,
backyard baseball
in the middle
of February.

Robins grow fat
with this new warmth,
raking worms from asphalt
as rain pushes
more from the ground,
a form of terrestrial
waterboarding.

Some birds we've never seen
are finding their way North,
something missing in their migration—
policy, planning, the time it takes
to figure out what really belongs
and where.

In the highest places, stone
stumbles upon itself, slabs
askew. Some say we unfold
when we fail. Do they mean
we come undone, or blossom?
I hope that when we're at our worst
we'll flower, petal upon petal,
exposed so that others might see
the lengths to which we go
to hide.

Soon we'll be moving
North, like the birds,
across lines men draw
on maps, as if osprey
or tundra swan
could pledge allegiance
to Canada or the U.S.A.

You have to go
where the growing season
works, where there's enough
food and quiet
for your children.

I wonder how far
we'll travel
to find the place
where we belong,
if I'll finally see
the long stretches
of open land, sun
shining for months
on end, then disappearing
into a cold night
for even longer.

## Be Prepared to Evacuate

Tara L. Masih, *Massachusetts*

My simple New England ranch home was built in the 1940s as a summer cottage. Although it does sit over an underground stream that runs beneath the cement foundation and exits into a culvert across the street, into a neighbor's side yard, the house is not in the actual path of any rivers that overflow. Nevertheless, human progress in the form of a cul-de-sac erected on the hill above my backyard, made up of large modern houses with blank windows that overlook the smaller homes beneath them, brought the problems. Developers and landscapers gave no thought to the repercussions of removing topsoil and undergrowth and trees that absorb rainfall and slow down runoff.

Still, at the start, this construction and the subsequent run-off weren't enough to cause more than a small trickle of water from three corners of my basement when the water table was high and the ground oversaturated from spring meltings. The early cottage builders, who understood nature and knew how to work with it, had done their job well—a drain hole in my basement's center, at a lower pitch than the three corners, took the water that naturally flowed there. During rainstorms, my neighbors' newer homes filled with inches—sometimes feet—of water that had to be pumped. My old-fashioned drain did its trick, while I slept peacefully.

That's no longer the case. It strikes me that anyone who experiences the direct effects of global warming doesn't doubt its imminence. Farmers and gardeners who work the soil and have generations of knowledge in observing plant life know that the seasons are changing, that the horticulture is confused, that crops are not being yielded in the manner to which we've grown accustomed. Alaskans are watching sea levels rise, which threaten to drown coastal towns that have existed for centuries. Recently, I read a purportedly scientific paper claiming that overall sea levels have actually fallen, due to evaporation over the Indian Ocean;

I wonder what that scientist would say to the Alaskans who are watching their homes float away.

On a smaller scale, I've watched the water come faster, rise higher, and invade more frequently every year that I have lived here. This year, in early 2008, I dread the satellite images that predict heavy spring rainfall. I cringe at the sound of punishing rain on the roof, and run down to the basement constantly to see if the water spilling over the congested gutters is migrating beneath the topsoil that can't hold its grip on it any longer. We are getting monsoon-like rain now in the States. These are not the fast-moving thunderstorms I grew up with, the kind that passed quickly, the kind after which you could go out barefoot and pretend to fish in the little puddles and running streams of clear rainwater they left behind.

These storms are slow moving and sit over your home for hours, days, seemingly endless, causing floods that are inevitably contaminated with sewage and chemicals. The kinds of rains that force animals from their underground burrows and keep pets from finding their way back home. After the last flood, we had possums temporarily move in under our tool shed and a cat take refuge on our back porch for a couple of days before the water subsided. When I checked in on a neighbor whose husband was a firefighter and never came home during these crises, she told me stories of people he had to rescue from stranded cars, and of homes he had to enter, full of water and raw sewage floating around. "We decided," she said seriously, "that if that ever happened to us, if the sewage ever backed up into this little house, we would just get in the car and drive away. Never look back."

If you've never had to do this, you don't know what it's like to keep a basement from flooding when the water is rushing in so hard that the drain fails. It's the end of winter, so the water is still icy, the rain painfully cold. It always seems to get worse as the sun goes down and the water is at its coldest. While you try to do the basic things like eat and squeeze in a final shower (in case you lose electricity and hot water), you have to monitor the water level. Every half hour you go down the basement stairs for a look around. When you see the drain failing and the water suddenly rising (it never happens gradually), it's a fight with the water's schedule to save your utilities and set up the pump. The gas furnace, washer and dryer, and gas water heater are all up on blocks that used to be high enough, but not anymore. You lost your water heater two years ago in the flood that brought FEMA to your town. Outside, you hear the fire trucks helping neighbors, who have learned how better to cope when floods threaten, to shut off their furnaces and electricity and to move in with a relative. You have nowhere to go, so have to win this battle with the water. You find the pump and fight with the pieces in desperation and anxiety, with that damn inflexible black hose that's like a stiff, stubborn, ornery octopus. You're standing up to your ankles in ice water, the plastic grocery bags tied around your non-waterproof boots failing. Water is seeping into your boots, and the hose refuses to cooperate. You finally get it hooked up and somewhat straightened and take down a window screen and open the window to the searing blast outside. You hope no one will be out taking advantage of this weather, looking for easy access into homes. You have no choice but to leave yourself and your son vulnerable while you exit the hose to the street and leave the window open for days. Now you have to turn on the pump without electrocuting yourself. You are dazed and tired and it's easy to make a mistake and do things in the wrong order. You did once and were lucky to have caught yourself before inserting the plug. Once that's done, you go out in the cold rain with your wet boots and frosty feet and work to set up the hose so it flows properly into the street and eventually to the storm drain. It's cold. Did you say that already? It's dark, but for the little illumination the street lights give off behind the screen of pelting rain. You curse being a single woman and have to knock on a neighbor's door at eleven at night. He answers in his wet gear; he is waging his own personal battle with his basement and, after a brief attempt to help you, has to rush back. Out of desperation, you drag your ten-year-old son from his heavy youthful sleep, get him dressed and into a raincoat and snow boots, and send him out to the street while you wrestle with the pump and exit hose, yelling into and out from the storm, back and forth, "Is it working?" "No!" This goes on while the water is an inch from putting out the gas water heater. Finally, blessedly, the answer is "Yes!"

This is not the end of the story. It goes on. You can't sleep while a stream runs beneath your bed. You can hear it, invading. Rising every hour in a state of disoriented half-wakefulness, you make sure the pump is working. Cold wet boots turn colder, as your body temperature plummets in the night. The hose will stop working several times, and you will have to make the trek back and forth yourself between the basement and the exit hose till you are screaming at the heavens to STOP. Stop making so much rain, stop it from coming, from invading your home and your sleep and your life. It feels as if that whole chunk of Indian Ocean that supposedly evaporated into the atmosphere has migrated west and dumped its vast expanse of water on your town. It feels like it is raining an ocean.

This goes on for three days. You shiver with cold and sleeplessness and ragged nerves for three days. You yell instead of speak to your anxious son. You have nothing to give him right now, no patience for anything. The pump works till it clogs with debris and dies, and then you resort to sweeping the water to the drain that is starting to recover. You sweep the water every twenty minutes or so, but as you sweep, more just comes in. You can't get anyone to come dry up your home. All the services are booked and the priority is to the larger homes that will pay more. You're on your own. A friend from New Orleans tells you to use fans and white vinegar to get rid of the smell. The fans work. You throw out the waterlogged, stinking boots.

A few weeks later, the torrential rains come again. And again.

I've heard that in a mere six inches of rapidly moving water, you can be swept away. And that flood deaths tend to occur because people try to outrun them, rather than move to higher ground. I can't fathom being in the path of a tsunami, or of a Katrina storm surge. Or dealing with flooding like that which took place a few weeks after mine, in March 2008, spreading across the nation's heartland from Texas to Ohio, sending all the rivers over their banks by record levels for hundreds of square miles. I've seen our parks and roads and intersections impassable, but there has always been some route out, some open road that allows for escape to higher ground. For a week, we had to go to another town for food, because our town's supermarket was flooded. We felt odd, out of place in just the next town over.

Yet this is nothing. Nothing compared to what the folks in New Orleans experienced, and are still dealing with. That flood left environmental refugees by the score. Still living in trailers insulated with material emitting formaldehyde, they are now contaminated by the very walls made to provide refuge. There are homeless people with jobs, living in parks and under highways. Yes, they have jobs but are still unable to afford rent or housing. They are there, as I write, in the shadow of the mayor's office. People like you and me.

I still have a home. But as global warming continues and these extreme weather patterns increase, besides first worrying about our planet and wildlife and our physical beings, I worry that our economy will never be the same again. I'm sure that part of the housing collapse and bank failures has to do with the fallout from the insurance payoffs to Katrina applicants. Millions of dollars have had to go to cleaning up disaster after disaster—hurricanes, tornadoes, tsunamis, earthquakes—and millions will be needed for future ones. Countries—people—cannot be productive when they are working at simple survival (one reason the Western world is ahead of Third World countries). Even food prices are skyrocketing because of the rain's effect on crops—flour is up four times in price, because wheat crops are poor, causing a trickle-down effect on the prices of baked goods and driving small bakers out of business. Maybe the economy is what will eventually wake up lethargic politicians, rather than images of drowning polar bear pups or human beings. Maybe as more lower- and middle-class Americans' lifestyles become like those of Third World citizens, we will speak up.

Everything is connected.

It's not enough to stand at the edge of a precipice or canyon and watch a mountain stream roar its way down to flatland to be able to understand the powerful, incessant nature of water. You can only conceptualize it through sight and auditory impact. To understand the destruction of water and to finally feel the powerlessness of human beings to stop its obsession with reaching lower plains, you have to be in its path. You have to be trying to stop it or to outrun it to recognize—you can't.

## Weather Weirding, 2012

Barbara Crooker, *Pennsylvania*

An odd spring, crocuses blooming
alongside tulips. Bluebirds, confused,
go back to sky. In March alone, fifteen
thousand heat records broken. Each breeze
wafts apocalypse. Colorado blazes
up in flames. July's transposed
with August; peaches and corn
a full month early. A string of days
over 100°, a rope of malignant pearls.
Katydids kick up their cadence, *did not,*
*did too* in the heat of the furnace.
This used to mean six weeks to frost;
not now. The Des Moines River
reaches 97°, thousands of fish
turn belly up. The Mississippi shrinks,
loses shipping lanes. Sea water's
too warm for nuclear power. Two months
early, chrysanthemums explode in August—
blood red, chrome yellow, rust orange.
And then, in October, barreling up
from the tropics, mating with a nor'easter:
Sandy. And what comes next?

*The first of this pair of poems about Superstorm Sandy is shaped by the author's experiences as a child in Hungary during World War II.*

## A Guest in the House

Paul Sohar, *New Jersey*

### *Hurricane Bombs*

world war II without sirens
bombs without planes
just tons of raw timber sent crashing down around us
exploding into splintered pulp and crippled branches

the hurricane smashes the night
into ruins of sound
tanks trotting over the roof

artillery banging away
on the black-&-white keyboard
of power failure
lit up by the crimson roses of emergency lights

except there's no war to win
no battle to lose
no good guys or bad
(are there?)

even back then
they kept switching sides
and where are they now

where are the battles they fought
the wars they won

mile-high tulip poplars bomb the roof
and here I am without an air-raid shelter
without a siren to listen for

even the wind has lost its direction
how can I tell where is the hurricane
and where are the liberators
or whether I am a grownup or still a child reaching for
the hand of my guardian angel

## *Powerless*

During the power outage I am
a homeless person camping out in my own home,
dispossessed by all appliances,
bundled up in rags from the bottom
of the utility closet

I sit in a corner begging for heat and light
and water to wash the bitter face of my soul,
begging my home to put its arms around me
and make me feel at home

Give me a cup of warm tea at least
before my words freeze too;
the word electricity already feels
like an icicle in my mouth and I'm ready

to take all these dead gadgets with their green
eyes tightly shut and pile them up in the living
room for a big bonfire. Let the flames
lift the house out of its indifference!

## The Things We Say When We Say Goodbye

Alan Davis, *Minnesota*

After Hurricane Katrina, my uncle Joe's apartment house in the Lakeview neighborhood of New Orleans had a spray-painted scrawl on it, indicating that it had been checked for bodies and none had been found. But his body was inside. A volunteer re-checked the apartment as a favor to my cousin and discovered Joe's remains. He was eighty-six. Apparently whoever had "cleared" the house had missed the body, or perhaps they hadn't searched the building at all.

Ever since he'd retired at age sixty-five from his blue-collar job on the docks, Joe had taken classes at a college in New Orleans, because he had missed out on an education when he was young. The year he died, he was a full-time student. The last time I'd talked to him, he'd wanted to discuss a term paper he was writing about Walker Percy's novel *The Moviegoer*. I knew the book, about a New Orleans that is now extinct, washed away by the flood. The protagonist, Binx Bolling, is a stockbroker who takes a leap of faith: "There is only one thing I can do: listen to people, see how they stick themselves into the world, hand them along a ways in their dark journey and be handed along, and for good and selfish reasons." I have never encountered a better philosophy of how to live.

"Do you think our life journey is vertical or horizontal?" Uncle Joe asked me, referring to Percy's idea of the vertical search for meaning within oneself versus the horizontal search for meaning in the world.

"I think it's both," I said, "vertical and horizontal."

"Yes," he said in his careful fashion, "that's what I think, too." He had a genteel New Orleans accent, as if he'd come from a monied district of cultivated gardens and carefully pruned banana trees and not from the Irish Channel, where not only Irish but also German, African, and Italian Americans, like Joe, lived.

"You know the Jesus prayer?" Joe asked me.

My uncle appreciated religion and literature and was able to hold in his mind both Keats' idea of negative capability ("When someone is capable of being in uncertainties, mysteries, doubts, without any irritable reaching after fact and reason") and the Jesus prayer ("Lord Jesus Christ, Son of God, have mercy upon me, a sinner"). I think those beliefs, in negative capability and in prayer, helped him keep his peace in the face of trouble, but like Binx Bolling in *The Moviegoer*, he often refused to discuss what he believed.

One afternoon when I was a boy, Joe and I were walking along the levee overlooking Lake Pontchartrain, and a man came out of nowhere, clearly haunted by something that he couldn't put behind him. (You meet a lot of people like this in New Orleans.) He yelled in my uncle's face about the Bible, waving his arms wildly, as if to some unheard music. "Listen to this!" he kept shouting. "It's the *truth*. I have the truth and you don't get it. You just don't get it!"

My quick-tempered father would have told the guy to "get lost, you filthy animal." But Joe let the man have his say, then turned and stared out at the lake, which was as placid as a bathtub that cloudy afternoon. He said to me—I must have been twelve at the time—"You ever notice how some people try to make it their business to pass judgment on other people's business?" I didn't know how to reply, of course, but these days his words come back to me all too often.

"Yes," I said to Joe the last time I spoke with him, "I know the Jesus prayer." I'm a recovering Catholic and was brought up in southern Louisiana; of course I know it.

"I still find myself using it sometimes," he said. "What do you make of that?"

"Does it help?" I asked.

"I don't know," he said. "It's like sneezing. I just do it." His voice quavered a little, the way my mother's handwriting did in her letters toward the end.

"Well, then," I said. Those were probably my last words to him, aside from the things we say when we say goodbye.

Joe and I spoke infrequently and seldom at length, but he was often in my thoughts. He had respect for me because I, too, loved books, and he worried about me the way an uncle does about a favorite nephew. He was a pacifist, a slightly built man who always smiled with eyes wide, as if life could still surprise him. He was a wayfarer, making his journeys on foot or by bicycle, living alone in a simple apartment with his books and his thoughts, keeping his money under his mattress, always trying to make sense of things.

After my mother died, six years before, Joe hitchhiked from New Orleans to Lafayette for her funeral but ended up at the wrong funeral home. A mourner there gave Joe a ride to the right place, a sprawling facility with space enough for six viewings. He hobbled in and waved to me. He had on a dark crew-neck sweater, slacks that still held their crease, and a patched sport coat that had seen better days. "Where's Dottie?" he asked me. That was my mother, his sister. They'd been very close.

The last time I saw him he was on his bike, which wobbled a little because his legs had lost some of their strength, and he waved to me over one shoulder as he bicycled away in the direction of Lake Pontchartrain. It was a bright day, and the white sails in the nearby marina fluttered against a blue sky. I was waiting to catch the city bus back to the French Quarter, where I'd meet my sister near the levee for a bowl of gumbo and a beer. I stood there with a backpack over my shoulder, wishing that I had a bicycle too.

Uncle Joe drowned in his apartment. I like to think that the Jesus prayer helped him as he died—or maybe Keats, or something else he'd come across in his reading or his wayfaring.

There was a point, during the disaster, when everybody thought that the hurricane had passed and that the worst was over. Then the levees broke—not from storm surge, engineers now think, but because the soil beneath the concrete walls was too weak. Nobody was there to help when the water started rising—a foot a minute in some places, I've been told.

Joe might well have been asleep when he felt Lake Pontchartrain lapping at his bed. I don't know. There would have been no electricity, but maybe he had a flashlight that he turned on and beamed around his room. It couldn't have taken him long to gauge his chances: he'd recently broken his hip and had limited mobility. Perhaps he was a bit dazed from painkillers. It was too late to crawl across the flooded floor. *My horizontal life is over*, he might have thought. *My vertical life is about to begin.*

Maybe then he recited the Jesus prayer.

I like to think he didn't flail and cry out, both because it's too painful to imagine and because he wasn't that kind of man. I prefer to think he floated as long as he could, and then, when there was no more air to take into his lungs, he closed his eyes and breathed in water. There was a sense of suffocation and panic, followed by peace.

It could have, and probably did, happen otherwise. I expect he wasn't ready to die—who is?—and fought against his fate, despite his age, despite the Jesus prayer, despite his almost Buddhist belief (though he would never have called it that) in the way each moment can be an eternity if we approach it without expectation or regret. Yet I like to think of him facing death as he did life: with dignity. Perhaps he wondered if he would be seeing his sister, my mother, again. She was a devout Catholic until the very end, when she died with six of her seven kids gathered around her oxygen tent. I don't know if Joe attended Mass on Sundays; it wasn't something the two of us talked about. But I believe that, like me, he had an unspoken faith in New Orleans, the dishabille city of our dreams and nightmares, city of Lent and Mardi Gras, of discipline and dissipation. I loved that city, where I was brought up and where Joe lived for eighty-four years, and I had faith that, wherever I went in the world, I could always return to its dilapidated glory.

I see now that disaster was always, is always, a heartbeat away.

CHAPTER TWO

# Species Out of Joint

## Search

Margarita Engle, *California*

Eggs of the sea turtle
are male in cool years,
female in warm—

no sex chromosomes
to keep a balance

as heated oceans fill
with wandering spinsters

searching
for mates.

## To Wit, to Woo

Kathryn Miles, *Maine*

The barred owl arrived in the full light of morning, silently landing on an ash branch at the edge of my yard. There she sat for the better part of a day, draping belly feathers over her otherwise bare feet and closing her eyes against the shrill wind. I might have overlooked her entirely, had I not risen from my desk to answer the phone or make another cup of tea or whatever else it was that caused me to stir at precisely that time, and thus to catch the settling movements of this raptor. I was captivated.

Owls seem uniquely suited to prompt this response in us. As I sat on my kitchen table, watching this one fill the canopy of the ash tree, my mind flitted from a sense of immediate wonder to the complex and sometimes contradictory mythic associations that an owl evokes when it lands in the company of humans. I thought about Thoreau's barbed celebration of owls as harbingers of a nature that we are inclined to miss, as "suggesting a vast and undeveloped nature which men have not recognized. They represent the stark twilight and unsatisfied thoughts which all have." I recalled that Athena—the Greek goddess of wisdom and justice—bore the symbol of an owl, as did Lilith, the dreaded Sumerian goddess of death. And that, more recently, women from Brittany and Saxony believed that owls bring fertility and healthy childbirth, while their counterparts in Morocco and Malaysia alleged that owls kill newborns and steal the souls of children. Even today, many Cameroonians call owls "the bird that brings fear," in contrast to those in Indonesia and Japan who insist that an owl call remains the best predictor of when it is safe to venture outside.

It must be tough business, maintaining so many mythic associations in so many places. Throughout that first morning the barred owl sat in my tree, I thought about these contradictions and more, wondering how such a quiet animal manages to shoulder them against an unforgiving wind.

Later in the day, though, another thought surfaced, more immediate and disturbing. *What was this owl doing in the daylight?*

For all the disparity in the collective mythology of owls, these stories share one very important quality: they are usually set at night. Over 125 owl species exist worldwide, and the vast majority of them are nocturnal; of the twenty types of owls present in North America, only five are considered diurnal, or daytime, hunters. Barred owls, like the one in my tree, function best under the cover of almost absolute darkness, where their keen eyesight and nuanced auditory system offer them great advantages over light-dependent daytime dwellers. This attribute made me all the more intent on monitoring my unexpected morning visitor, and so I watched for hours that first day, worrying about what her presence might mean.

Admittedly, barred owls are not uncommon in the wooded foothills of Maine, the place I currently call home. To the human eye, this is a liminal landscape: a daytrip away from both the rocky coast celebrated by Rachel Carson and the northern terminus of the Appalachian Mountains, where John Burroughs and countless other nature writers sought inspiration. Unlike those more iconic places, my landscape is one of intersection and contrasts, where hardscrabble farming and a waning timber industry meet emerging enterprises like biotechnical labs and guided hunting tours. Life here tends to be scruffy, bearing little resemblance to the New England featured in glossy brochures and guidebooks. Nevertheless, it is a habitat that members of the *Strigidae* family seem to love, what with its savage swamps, mixed-coniferous woods, and vole-thick meadows. In fact, this place is what scientists at the Maine Audubon Society like to call a "hot spot" for owls, and species like the barred have flourished here since long before any human arrived.

When my husband and I moved to Maine in 2001, we were captivated by the overwhelming presence of owls. Lying in bed at night, we'd hear the cry—rendered by Shakespeare as "tu-whit, tu-woo"—rise up and out of the dark woods. As we listened, I had no difficulty understanding why generations of people across the globe have associated such calls with the underworld: it seemed such a ghastly voice to my untrained ears. Since then, I've grown to love that eerie crescendo piercing the still blanket of night, and I happily side with Thoreau in thinking that this cry somehow heralds a reminder that there is a nature that we do not recognize or even acknowledge. Perhaps this dark unknowability has been why the mythic role of the owl continues to woo us with the implicit wit in its call. Perhaps it is why the literal owl has remained a welcome and comfortable mystery for us, occupying as it does those dark and unknown places that we imagine but never see. On many a call-filled night, I certainly have thought so.

That, I think, is why I found my quiet daytime visitor so troubling. She should have been unnoticeably absent—a disembodied voice calling out of the wilderness of night, or at most a hidden form napping in the cavern of a dead tree or a church steeple somewhere, protected from the elements and waiting out the light until it was time for her shift to begin. Nevertheless, she stayed on that ash branch for hours, still as a secret and unprovoked by the nuthatches sharing her tree or by the curious human creeping across the yard for a better look. She remained there, day after day, at the edge of my yard. Her continued presence signaling, with its lit silence, that something in the world is amiss.

What is that something?

Perhaps the most unsettling truth of all is that no one knows for certain. While we may love that which is unknowable about owls in story, most ornithologists say that mystery can be frustrating as hell when it comes to science. They do know that the diurnal appearance of an owl usually indicates that the bird is food-stressed. They also say that a lot of these hungry birds have been making appearances throughout the northern United States lately. In recent years, avian listservs and blogs have been humming with reports of daylight sightings of owls like the barred, as well as the surprise appearances of great gray, snowy, and boreal owls—three species normally found far north of the continental U.S.

Biologists have taken notice. They say rehabilitation centers are filled beyond capacity with sick and injured owls. Reports of dead owls found on roadsides are rising as well. Accidents between raptors and vehicles aren't all that unusual in and of themselves, but the growing frequency of these incidents is. So, too, is the rising number of owls breaking the boundaries of what we consider normal habitat and behavior—like foraging at a Dunkin' Donuts dumpster in Burlington, Vermont, nesting on the fields of Boston's Logan Airport, or dropping down from Canada to hang out in places like downtown Missoula and Seattle. In 2005, an unprecedented 5,000 great gray owls—North America's largest and most mysterious *Strigidae*—moved from Central Canada to Northern Minnesota. This unexpected arrival brought at least as many birders, who traveled from as far as Eastern Europe to observe these raptors. Some were content to take photos of the great grays; others, however, came to ask what the presence of these normally boreal birds might mean.

Similar questions are being asked where I live, too. Earlier this season, a great gray appeared on the edge of our town and claimed a nondescript power line as his perch. There he, like my barred, sat day after day, patiently enduring the gaze of birders and curiosity seekers who drove down the otherwise lonely gravel road to catch a look. Like me, they were spellbound, enraptured—by mythology, by life lists, by the cognitive dissonance created when anything, creature or otherwise, is utterly out of place. It's a difficult and complex thing to stand, gazing upwards, and scrutinize an animal who, distressed, was forced to cross so many natural borders. Yet, in some real ways it seemed as if all we could do—or rather, what we *needed* to do—was bear witness.

This, of course, is a human conceit, and one that risks anthropomorphism in some serious ways. Wild animals don't want witness. Our local bird expert says he thinks the owl came here because it wanted to die alone; that assumption probably butts up against anthropomorphism, too, but perhaps gets closer to the owl's instinctive response.

In the end, it's what the owl got. A week after the great gray arrived, he was found dead on the ground near his perch. As far as we know, no one was there when he died, since it was at least a day or two before his body was discovered. State wildlife experts conducted a post-mortem study, hoping to determine what brought this visitor across several discrete ecosystems and eventually caused its death. Their findings were sobering.

As best as the biologists can tell, the owl died precisely because it crossed ecological borders. More specifically, it died of an infection known as aspergillosis, which is caused by a fungus commonly found on dead leaves and dried grains, as well as compost piles and bird feces. In humans, the condition can cause an acute respiratory condition for people with asthma or compromised immune systems, but most of us fend off the fungus with no noticeable effects, as do most resident predatory birds. Those owls coming down from northern territories, however, are far more susceptible to the disease, since their lungs haven't acclimated to the presence of the fungus.

Normally that's not a problem, since owls prefer to stay put: a predilection that makes the current movement of these raptors all the more beguiling—and dangerous. Most frustrating of all, no one knows for certain why these owls are now leaving their own landscape for potentially compromising ones.

For if anything, the presence of these raptors seems to signal the very real boundaries of modern scientific knowledge, the limitations of our standard ways of talking about avian behavior. For example, migration—that bridge linking summer and winter habitats for most birds—usually exhibits great regularity in terms of specific habitat, path taken, and moment of travel. Not so, however, with these shifting relocations of barred and great gray owls, which seem more akin to avian irruptions, the unexpected and irregular movement of a large population *en masse* from one landscape to another.

There are many reasons why these mass movements occur, but in the case of owl relocations, global climate change and human intervention are probably to blame. Most northern owls prefer to maintain territories deep in the boreal forest, the massive swath of canopied land covering over 1.3 billion acres in central Canada, once an untouched haven of mature needleleaf trees. A rapid increase in the number of forest fires over the past decade, coupled with increased logging pressures, has changed all of that, causing the collapse of food sources for northern owls and thus pushing them into my region. This same irruption theory may explain why other northern owls are appearing in densely populated downtown cities like Minneapolis and Chicago, and why a barred owl now spends her days in my tree: they are competing with too many other owls during nighttime hunting. The added population pressure has exceeded the carrying capacity of the landscape, and the owls are hungry.

This hypothesis makes a good amount of sense, but so far science hasn't been able to find much conclusive about the owl irruption phenomenon—except that it is increasing. Meanwhile, much of what scholars thought they knew about the range of owls is changing as well. With or without the increased forest fires, climate change is beginning to redistribute territorial ranges for a variety of species, and that is making it increasingly difficult to classify basic population shifts in a scientifically meaningful way. Much of this difficulty comes down to shifting notions of borders and the spaces they separate.

Take my backyard visitor. If you draw the number *7* across North America, beginning with a horizontal bar across Southern Canada, and then a thick diagonal slash cascading from Nova Scotia down to the edge of Louisiana, you'll have a pretty good sense of the barred owl's range. That wide, zigzagged territory makes discussing latitudinal shifts of the species difficult. In its strictest sense, the term "irruption" applies to the appearance of a species in a biome where it is not normally found, but barreds have been calling my landscape home for centuries. We don't have a word for what it means when there are suddenly a lot more of them in this area. Or what to call it when they begin to push further westward, taking over territories normally held by their close cousin, the spotted owl.

In fact, we don't have words for many of the things happening to North American owls right now. Even when you commit to studying them, owls are maddeningly elusive, scientifically speaking. Masters of camouflage, these birds of prey are difficult to locate during the day, when their cryptic coloration mimics the texture of tree bark and limbs. At night, they are adept at nearly soundless flight, thanks to a unique layering of feathers on their wings. Both factors make tracking and observing owls downright difficult—even for trained experts. To make matters worse, wide variances in natural populations from year to year complicate any attempt to parse out trends in total owl numbers once you find them. If you add the fact that most formal ornithological research is restricted to a single bioregion (thus precluding a thorough understanding of an irruption from one region to another), you have a recipe for spotty knowledge at best.

Herein lies the greatest irony of the twenty-first-century owl: although much loved by amateur birders and pop culture aficionados alike, these birds have been tremendously difficult to study. As a result, we know a lot more about their place in comparative mythology than we do in place-based ecology.

This is a bitter pill for me to swallow. Like a religious novice, I find great solace in the sanctity of science: its elegant simplicity, austere language, and perhaps most important, its implicit belief in the knowability of the natural world. Linnaean taxonomy provides the rubric through which I have come to know the trees and animals in my forest. The principles of ecological niche theory explain the degree to which I might live within—or take from—that locale. I'd like to believe that these scientific concepts not only help me understand my place within this little biome, but also give me the tools I need to make it a sustainable one. How can it be that biology cannot explain the place of owls in this same landscape?

The answers are as varied as the symbolic representations of owls across the globe. When I contact Susan Gallo, a wildlife biologist for the Maine Audubon Society, she speculates with me about possible causes. One reason for our lack of uniform answers, she says, might stem from our all-too-human suppositions about landscape. Gallo says that she and other researchers have always known that they lacked crucial information about the status of owl populations, but for a long time, committing to a formal investigation seemed too daunting. It was, she admits, a kind of statistical inertia across the country.

According to Gallo, ornithologists within any given region knew that they needed to be studying owls, but they couldn't build momentum in the scientific community for a collective survey. She adds that, in the past, this lack of initiative didn't seem like a problem; for most state wildlife departments, the decision not to track owls was based on a long-held assumption that most endemic owl species were both common and secure. Now, though, the increased sightings and mortality rates are forcing these same scientists to wonder if their assessments remain valid.

Luckily, says Gallo, owls do offer one advantage: they are among the most vocal of all birds. The usual accoutrements of bird communication, like showy plumages or elaborate mating dances, don't do much good in the dark, so owls have adapted verbal ways to telegraph these messages. They use that haunting "tu-whit, tu-woo" to vocalize territorial concerns, invitations to mate, or as a constant audio transponder announcing their presence to anyone who might be interested. Recording these solitary voices in the night may be the first step towards solving the emerging owl population crisis. And so, across North America, groups like Bird Studies Canada and U.S. Audubon chapters are launching new owl monitoring programs intended to glean a better understanding of how many owls exist where.

I find a kind of salve in knowing that this research is happening. I've been watching the barred owl in my ash tree for a week now. Each morning, she arrives without a sound and sits well past when I've had my lunch, while she presumably looks for her own. I want to know more about her and other owls. I want to hear them once again during their nocturnal hours and understand what little information conservation groups are gathering. I beg permission to join a group of monitoring volunteers. Happily, Gallo agrees.

Later in the week, I join resident ornithologist Dave Potter and nine volunteers rendezvousing in a frosty parking lot just after midnight. The volunteers look half-sleepy, half-wary in the dark of night. Most stay in their cars, staring at an invisible point somewhere on the horizon. Others lean against the van we'll be taking, looking like they appreciate the security of a boxy, lit vehicle. Potter laughs at our timidity and reminds us that we're on owl time now—which means we'll probably be out until 4:00 or 5:00 a.m., depending upon the whims of the nocturnal birds. It is surprisingly cold (about 23° F by Potter's count) and disappointingly windy. I wish I had one more layer on top of the five or so I already wear. Potter, who sports four brightly colored flannel shirts and a brand-new stocking cap, seems undeterred by these conditions.

Tonight, Potter's accomplices range in age from their late teens to early forties, though it's hard to tell with all their cold-weather gear. A few admit they are Potter's biology students and hoping to garner extra credit, or at least their professor's favor. All are avid birders and repeat participants in the study. Once inside the van, they laze around with comfortable experience and a vernacular to match. These birders love rhyme: they refer to the count as an "owl prowl," to themselves as "bird nerds." Eight of the ten volunteers are women, which Potter says is a typical demographic for his research expeditions. When asked, he says he would prefer not to speculate why. The women look relieved.

By way of orientation, we lean towards the driver's seat of the van while Potter plays the standard CD issued by Audubon.

The disk begins with several minutes of silence (time to assess ambient noise). This dead time—which we will learn to rue while standing in the pre-dawn chill several hours later—is followed by a series of three owl calls: long-eared, barred, and finally the great horned.

As we listen to the thirteen-minute recording, the other volunteers identify all of the owl calls before Potter announces them. Miraculously, they can also make out the sounds of Canada geese, nighthawks, and other seemingly imperceptible calls comprising the background noise on this recording. I am impressed, and deeply intimidated; I say as much, and get the impression that this pleases my new friends.

As we drive to our first of ten stops, Potter explains that, when we return sometime just before dawn, he will feed our information into a new international database on raptor populations. The data-point provided by our inky time outside may help to answer the question of how climate change is changing avian populations—and the places these birds call home. He also hopes it will tear down some of the boundaries science has erected for itself.

"The question of owl movements isn't going to be answered by researchers and scholars. The long-held belief that only they can solve natural conundrums is an outmoded myth of the most dangerous kind," he says. "The only way we're going to get answers is by engaging the public in citizen-based observations. Not by relying on hotshot researchers. The sooner we get rid of those distinctions, the sooner we can learn something."

I tell him about my own observations of the barred owl who has been spending her days in my ash tree. Potter nods silently, then says that he has a barred owl in his yard, too. His, though, is smaller than the shape I approximate with my outstretched hands; based on that guesstimate, he thinks his barred owl is probably a male. Potter's wife, Lonnie, keeps track of the owl during the day, and he thinks the two have formed something of a bond. She's named the owl Ernest. Potter likes to start some of his classes with updates on Ernest to get students thinking about applications for the otherwise abstract theories they are learning.

This prompts commentary from the back of the van. All of our fellow volunteers, it seems, have something to say about on-the-ground inquiry. They rave about the experience of cataloguing these owls, in spite of the cold and the quizzical looks from passers-by and, on at least one occasion, a nearby homeowner who responded to their CD with a shotgun blast. Each year, they swear they will never do this again, but each year the mystery and the romance of this singular bird keeps them coming back for more.

The first of our stops takes us to a gravel road sandwiched between a demolition derby racetrack and an industrial composting facility. It is a place far from the mystical scenes described by the volunteers when they waxed poetic about depictions of owls. The remaining stops are similar: below flashing traffic lights, tucked inside a much-decorated cemetery, along an overpass near a highway. Even in a working landscape like ours, these are compromised, well-worn spaces—far more akin to the parking lots of Dunkin' Donuts and the tarmacs of major airports than the ideal habitat I had imagined for a nocturnal raptor.

I ask Potter about the nature of these stops.

"Mostly we pick places where it looks like an owl might appear," he says, "but that's just based on our best sense of things. The places where nature and civilization meet are the places where we have data. It's about access, really. We don't know what's happening deep in the woods."

He and other scientists are hoping that the composite results of surveys such as ours might begin to change that. Once completed, our statistics will be added to two new databases, one created by Bird Studies Canada and another housed at Cornell University. These web-based clearinghouses integrate the information gleaned from other surveys fledging across North America, and they represent the first organized attempt to think beyond local research, at least where owls are concerned. Minnesota and Wisconsin have started their own citizen-based survey, and organizations in Connecticut have begun investigations, too. In time, all of these programs will—or should, or might—provide the databases with enough information for scientists like Dave Potter and Susan Gallo to determine concrete trends in owl populations across the continent.

Getting that kind of statistical information is an exercise in patience and chance. At our first few stops, we volunteers lean into the wind, hoping to hear a call. A faint chirp in the background at our first stop elicits an audible gasp, followed by hopeful speculation that it could have been an owl. We really, *really* want it to be. I wonder aloud if this desire affects survey data; Potter shrugs and says he isn't too worried about that. He knows there are owls here—he just doesn't know how many.

At our third stop, the volunteers' hopes are fulfilled: a mournful barred owl somewhere just out of view joins our CD in a long duet of call and response. The sound has an effect no less than that of a strong electrical charge. It, and more distant calls, sustain us for the first several hours of our trek. Still, despite our collection of coffee and hot cocoa and snacks, we are a weary group by the sixth stop. Despite the dramatic rise in populations caused by the irruption, owl calls are sparse this year. By the seventh or eighth stop, a few volunteers opt to stay inside the van. By the ninth and tenth, not even the novelty of hearing a great horned owl can keep people lingering by the portable radio after the recording has ended. Our crew is ready to go home.

As we make the cold drive back to our waiting cars, I ask Potter about what we did—and did not—find on our trek. He speculates that we could be witnessing early signs of a crisis brought on by overpopulation and a subsequent lack of food. Then again, it could have been a quiet night for some reason we don't yet understand. In the end, he says, there is still too much that we don't know about the predicament of resident owls. One thing remains certain, he says: through their silence, these owls are trying to say something crucial about our environment.

I tell him about the sense of foreboding in Thoreau's depiction of owls in *Walden*, the idea that owls speak to a kind of dissatisfied netherworld, that their calls announce "a dismal and fitting day" and "a different race of creatures" about to emerge. Am I being melodramatic in thinking that we're proving this description right? Or that there was a prescience in ancient representations of owls as both wise and ominous?

Potter says he doesn't know—but we both agree on one thing: if nothing else, the dawn of a new century has unequivocally changed our discourse about not only this landscape and its inhabitants, but also the categories we use to understand it. For the first time, current generations are collectively acknowledging what some have long held true: our climate is changing. Along the way, we are redefining everything we thought we knew about the natural world, particularly when it comes to the systems we use to make sense of it. Too many of our stories and scientific theories lag behind this change. It takes examples like the plight of the barred owl to recognize as much. Maybe this acknowledgement is, in the words of Thoreau, both dismal and fitting. Maybe it takes something as silent as the daytime appearance of a hungry owl for us to heed that call and to answer it with one of our own. Or to shift our own sense of tidy boundaries and edges—especially where knowledge is concerned.

These are the thoughts that accompany me each subsequent morning as I look to my ash tree for a sign—or at least for that now familiar presence filling the canopy. I find neither. And yet I keep looking, torn between whether or not I want my owl to return and very much wanting to know what has become of her.

The weekend after our trip, Potter emails me with results from the survey. As he and other scientists suspected, the barred owl population appears to be in trouble. I tell him that my visitor, though once continually present for about ten days, has not returned in over a week. I considered playing the Audubon CD to see if I could woo her into responding, but I worry I might inadvertently send a message about food or territory that would complicate matters further. I still don't understand what these calls mean—at least, not what they mean to a fellow owl. I ask Potter if he thinks her absence makes my barred owl a casualty or a survivor. He says he doesn't know—that we may never know. In the meantime, though, Ernest remains vigilant on his perch at the edge of Potter's property—looking very much a mythic figure. Maybe, Potter speculates, this silent owl will eventually explain what we cannot.

## Winter Visions

Paul Sohar, *New Jersey*

This winter brings unquiet dreams
yet we can't help dreaming on
even while the bears remain awake;

just today, the fifth of January, I saw
a mama bear with her two cubs
browsing right beside Millbrook road
in Worthington State Park, New Jersey.

I never talk to bears but today I should've
stopped and asked them
what was keeping them awake,
what dreams had slashed the cobwebs of their sleep.

Maybe they know something
we too should know;
the visions of insomniacs are etched by lightning
into real clouds in a fragile sky.

Or else these three bears I saw were somnambulists
like us, who want to keep on dreaming
that this winter is as real as any other,
no better, no worse.

## Trees of Fire and Rust

Margaret Hammitt-McDonald, *Oregon*

The breath of the mountain sparkles in my lungs. From the forest comes the crisp perfume of pine sap, as rich and dry as the afternoon light counting its golden coins through the branches. The trail exults in its altitude, discouraging most hikers with its steepness, and so my spouse and I have the thimbleberries to ourselves. Over our shoulders, Mount Hood leans toward us in paradoxical grandeur and intimacy.

In this seemingly most pristine of places, though, the damage shouts for recognition.

In photographs taken twenty years ago, Mount Hood is swathed in snow year-round. This summer, though, the mountain's rocky shoulders are bare and gray. What should be majestic seems vulnerable, this soaring and empty rock. Remnants of snow crouch in fissures, not the blistering white of forbidding angels but the craven beige of snow about to die. And in the absence of snow, a new danger has emerged: along the tree line, the deep green of shade and secrets is punctuated with alien rust, glowing like fire in the afternoon light.

Half of the forest has succumbed to pine bark beetles. With the warming of the world, the beetles have expanded their range and consumed more trees, transforming their cool green to these scorched candles. When we walk among them, we find the needles fallen and the branches curled inward, like the talons of songbirds frozen in the snow with their feet in the air, clawing at the sky. These trees have rusted like metal left in the rain. When we stand atop a viewpoint, we see whole hillsides raped by rust. The fiery corpses of beetle-stripped trees advance upon the huddling green of the survivors.

In this remote place, it's usually easy to overlook human-inflicted scars upon the living world. But when I look down, eyes opened by the rusting trees, I can trace the aerial scars of power lines and the lacerations of logging roads. The mountains are quilted with clearcuts. In some places, traffic sounds from far below waft upward in mechanical echoes of wind.

My own hands, I realize, have wrought these stigmata on the land. We drove on one of those roads to get here. Those power lines light our home. The trees torn from the wombs of these forests have given us paper and furniture, as well as the skeleton of our house. We have invited those beetles to set pines afire in death.

Thimbleberries rust in my mouth. Their maple syrup taste is the blood of trees.

Those trees of flame, the bare back of the mountain, those clearcut wounds, along with a summer day and the knife air that the mountains wield—all of these voices call for me to change my relationship to this beautiful and burning world, to seek ways to return the snow to the mountain and to sing the green into the trees. Indifference is too costly. I cannot forget the sweet incrimination of scarlet that the bursting thimbleberries have left all over my fingers. I cannot refuse the call for change from the frail and awesome places that are the treasures of my heart.

## Burning to Zero

Carla A. Wise, *Oregon*

As fires tore through Yellowstone during the long, hot, dry summer of 1988, the fear on people's faces made the nightly news, a glowing orange light in the background. Well over a million acres burned before fall rain and snow finally extinguished the fires, a massive, terrifying display of nature's power. Yet when I visited Yellowstone the following summer, what I found was a huge surprise: a diverse and beautiful patchwork landscape, not the place of devastation I'd expected from the TV images. True, some areas were burned down to a blackened lifeless zero, but in other places, the fires had been quick and shallow ground fires, and flowers were already blanketing the ground—Wyoming paintbrush, shooting star, monkey-flower, penstemon—while charred but still-living trees stood above them. Whole other chunks of ground seemed to have been skipped altogether, completely untouched and showing no signs of burns at all. More than anything, the land looked refreshed.

Looking at the vast burned forests of central Oregon today, however, I can't find much beauty, only black charred death. Driving along the highway that winds up through the Cascades over Santiam Pass, looking up the slopes of Three-Fingered Jack and Mount Washington, the land doesn't look renewed; it looks devastated. Too many fires, named for their specific locations—Fossil Creek, GW, Black Crater, the B&B Complex, Shadow Lake, Rooster Rock, Pole Creek—have burned here recently, and not much is coming back.

Some people who live around here are mad—at the Forest Service, at environmentalists, at authorities in general—the way locals were mad at the government about Yellowstone all those years ago. Wildfires have been something to fight about in the West my whole life: people want to blame someone besides themselves. That's much easier than facing the new reality—that we are rapidly losing the power to control western wildfires at all.

In the fall of 1988, as the snows were putting out the Yellowstone fires, I was starting a master's program in natural resource management. I was taught that forests in the West evolved with fire, and that some species depend on periodic wildfires for regeneration. Professors told us of fire's benefits for forest ecosystems: fire removes debris, enriches the soil, and allows for forest renewal. Eighty years of fire suppression, they instructed, had inadvertently allowed dangerous fuel levels to build up in many western forests, altering the natural balance. The best solution would be a combination of controlled burns to reduce fuel loads and a "let it burn" policy for naturally ignited fires in uninhabited forests.

Outside of academia, though, many people in the tourist- and logging-dependent communities in the West disagreed. From their perspective, restrictions on logging and thinning were largely to blame for wildfires and their devastation. The solution they proposed was to log and thin western forests wherever possible, and to fight all wildfires aggressively from the start. Today, this same debate takes place over forest management here in Oregon and elsewhere in the West, inflamed by each new fire—and by our unfolding understanding of climate change, which was just beginning to be discussed when I finished my degree in 1990.

Our forests are burning hotter and longer as climate change tightens its grip. It's simple: the changing climate is causing longer, drier summers, and better conditions for catastrophic wildfires. A study published in *Science* in 2006 showed that wildfires in the West had increased fourfold between 1987 and 2003, compared to the sixteen previous years. This upsurge corresponded with an average temperature rise of 1.5 degrees Fahrenheit, a 78-day lengthening of the fire season, and over six times more land burned.

Since then, evidence for the link between climate change and worsening western wildfires has only gotten stronger. A 2012 analysis of four decades of Forest Service wildfire records found that nearly every western state has experienced a huge surge in wildfires in the last decade, linked to dramatic warming and earlier snowmelt. Today, fire season is two and a half months longer than it was forty years ago, the number of fires of all sizes has increased dramatically, and on average, wildfires burn twice as much land area each year as they did forty years ago.

Scientists also have documented the bark beetle's march northward and upward to devastate once-safe forests. Beetle-weakened forests burn hot and fast. Climate change has changed the rules of the fire game, and we have yet to fully grasp what that means.

Denial, my sister loves to say, is not just a river in Egypt. I can't see that any amount of burning, thinning, or managing is going to save these forests from fire in the new world we are making. Climate change is drying out forests, lengthening the fire season, increasing droughts, and worsening insect kills. Each year, more forest succumbs to the highest-intensity fires, leaving little life behind—what I think of as "burning to zero." It is a bitter pill, especially when you love these lands.

Several winters ago, I skied through a patch of ghostly forest near Santiam Pass, looking for clues. This land was in the heart of an especially hard-hit area of central Oregon, with views in all directions dominated by blackened matchstick forests. I love cross-country skiing, but the black tree skeletons against the white snow and gray sky felt bleak and sad that day. It was silent except for the faint noise of cars on the receding highway, and the black and white landscape was an open question. Who knows what the snowmelt will bring? How many springs might it take for a blanket of flowers to be waiting under the crusty frozen cover?

Ever since that day, each time I drive over Santiam Pass between our home in the Willamette Valley of western Oregon and our cabin in the high desert, I gaze out the car window, hoping to see flowers and rebirth, the bounty and lushness I remember from Yellowstone. Spring comes, and then summer, but the vast dead forests along the Highway 20 corridor remain, barren and ugly.

I am beginning to view these charred remains as a repeating reminder of something that we all must understand and find peace with in these times. Destruction is one of the main forces driving the universe, and these vast, blackened lands are part of a cycle. I remember the Hindu god Shiva, and I think perhaps he can help me accept the fires, the climate crisis, the time I live in.

In the Hindu system, the endless cycle of creation, preservation, and destruction is expressed in the supreme trinity of Brahma, Vishnu, and Shiva. Shiva, the destroyer of the world—terrible, merciless, powerful, good, and awesome—is responsible for change both in the form of death and destruction and in the positive sense of the shedding of old habits. You aren't required to find Shiva's destruction pretty; you must simply bow to his power and divinity, and wait for the rebirth that will inevitably follow. I need to get to know Shiva, to see these fires as a symbol of the destructive cycle we are entering. I might be able to view the blackened landscape as cleansing.

Over the last few years, I've spent many hours learning and writing about climate change, and this work has been cyclical as well. Sometimes I enjoy the work, but at other times—when I get too emotionally in touch with the likely future, the extent of what we have done and what we are in for, and the ever-shrinking likelihood we will act in time to prevent the worst consequences of what lies ahead—I get overwhelmed. In 2050, my daughter will be fifty, my imagined grandchildren (if they exist) may be starting out on their own, and—if we don't change course—our world will be entering chaos.

Facing the truth of these times requires strength. Sometimes, I have enough. Strength comes from a variety of sources: my work—personal, political, and professional—to help solve the climate crisis; time spent outside in forests and wildlands;

time with my family and friends; my desire to be a good and positive mom and spouse. A climate activist I deeply admire said, "Do something, do anything, just don't do nothing." The comfort in heeding her words has been profound. I gain strength from others, as writing allows me to learn about people who are doing amazing things on behalf of the future of the planet and of our children. And lately, some deep, irrational optimism in me watches for the regrowth in the devastated forests that I won't see in my lifetime, and waits.

*Palm oil can be used to produce biodiesel, thus reducing greenhouse gas emissions compared to petroleum-based fuels; but oil palm cultivation has complexities of its own, as explored in the following short story.*

## A Jungle for My Backyard

Golda Mowe, *Malaysia*

The longhouse resonates with the excited voices of men, women, children, and hunting dogs, for a Member of Parliament and an officer from the Agricultural Department will be visiting this afternoon. About the chief's gallery, five of the prettiest girls prance like birds of paradise in their costumes of dark skirts woven in patterns of curling hooks, and embossed tin headdresses with standing pins of drizzling petals. The combined odor of fried shoots, boiled pith, steamed rice, and roasted pork permeates every nook and cranny of the house.

Door after door of longhouse occupants have placed plastic soda bottles filled with rice wine in the center of their family gallery. Twenty-three doors equal twenty-three families, all with a mission to ply every adult who dares step into their portion of the house with a glass of homemade *tuak*. Unattached women turn their eyes away as I pass; I am no longer the acceptable bachelor I used to be. I stand behind my grandmother, as she and another lady stretch to hang a piece of ritual blanket over a blank wall.

"Where is the pig the MP has to spear, Grandma?"

"Don't be foolish. We are not barbaric pagans."

"But I thought it is normal practice for a VIP to spear a pig before he enters the house."

"The chief slaughtered a pig this morning. It is cooking over his fire right now."

I look around the busy common gallery. "Why are those girls dressed like that?"

She smiles. "Don't they look pretty? They will serve food and drinks to the VIPs."

I raise an eyebrow, for when my grandfather was still alive, only men would serve male visitors. I cannot resist asking, "Will they sit on their laps, too?"

Grandmother furrows her brow and shakes her head. "I just don't know why you are so mean-spirited. I have always told your father that he has not brought you up to be well-mannered."

"Why are those government people coming anyway?" I say, trying to change the subject.

Her face wrinkles into a smile. "Our MP has received approval from the government for an oil palm project on our land. He and the officer are coming to tell us how we can take part in the program."

My lungs cramp as I try to hold back a cough. "Oil palm? You do not plan to use your land for that, do you?"

"Of course I do. What else will I do with it?"

"I don't know, Grandma. Putting all your eggs in one basket doesn't look like a good idea to me."

"Well, it is better than not having a job, isn't it?"

I heave a sigh: our conversations always end like this. It is true, I am unemployed. My last job had been in a factory where I breathed eye-stinging air for two years. Day after day, my ex-colleagues and I had gone to the panel doctor complaining of chest pain and of coughs that wouldn't go away. Each time he would give us something to make it more tolerable—blue pills, green pills, or blue and green pills—but the pain always returned. One day, I noticed that the paper cut on my finger wouldn't heal. It made chills run up and down my spine. My friends and relatives tried to persuade me to stay, at least until I could find another job. They argued that it was a good company, for the workers were paid on time each month and there was an in-house panel doctor to take care of all our medical needs. But the air made me cough so much that my lungs had begun to hurt, and the doctor could not give me anything to take the chill away, so I left.

I have not felt that chill in a long time, but today as I watch the people hurry about their duties, the chill returns. Maybe I am catching a cold. A warm breeze whistles past the bamboo forest lining the front of the longhouse. From the distance comes the long-call of a red ape. I know that if I step out of the house and into the forest just twenty yards away, I will find a place that will make me feel small, yet significant. Insects will greet me from the shadows of leafy undergrowth where rattan vines will tempt me to stay with their thorns, and if I should venture deeper, a fig tree will nourish me with globes of fruit that hang down like a bead-curtain of green and red.

I watch the jungle sway and flutter, then turn my attention to old Uncle Gramong as he climbs the front steps with a large load of bamboo shoots on his back. He unstraps the basket, and wipes his sweat-drenched face on a sleeve. The shoots are still covered in their furry brown skin, and they will stay that way until he is ready to take them to market.

"Ooi, Uncle. That is a lot of *tubu*."

"Ya. Bamboos love the smell of smoke. Every time after I clear my rice fields for planting, these shoots will poke their little heads out of the ground."

"How much do you sell them for?"

"Two ringgit each."

"I will buy one from you, Uncle."

It is a good price. One head must weigh at least 1.5 kilograms. The old man will earn about twenty ringgit for his labor that morning. I stretch out my hand to receive the merchandise, careful not to rub my palm over the stinging hairs. Again the wind whistles, and a cramping sadness spreads across my chest. I look around; everything is in its place. I saunter back to my grandmother's family room with a peace offering in hand.

The important visitors arrive at 4:00 p.m. Melodious gongs and drums herald their presence to the trees, the birds, and the wind. After a thankfully short speech from the Member of Parliament, the Agricultural Officer stands up and tells the crowd thronging in the common gallery about the government's oil palm scheme. He speaks of poor farmers who became rich after joining the program. He shows them simple market data that help them visualize the money they will earn by being part of the scheme. Then he opens the floor for questions.

"What will happen to our land after we stop planting oil palm?" I ask.

The MP's penetrating gaze falls on me. "You don't look like someone from here. Where are you from?"

"I am from Sibu. I am visiting my grandmother today."

"What is your interest here?"

"Part of my grandmother's land will one day belong to me. I am asking to make sure that my family will not lose anything by joining this scheme."

The MP chuckles. "Your grandmother is still alive and you talk as though she is already dead." A disapproving murmur rises from the crowd. One of the serving girls passes me by without offering any refreshment.

The MP continues, "You have every right to ask. As you say, it is your land. Well, the land will be returned to you and you can use it for any future project that the government will come up with. After all, we have the people's welfare in our hearts. We want you to be successful, just like the other races in Malaysia."

Again, a fever-chill courses through my being and my tongue swells as though thick with a strange terror. I am angry with myself. I am angry because I let myself be insulted in front of my people. I decide to ask another question though I am cowering inside.

"I hear that oil palm needs a lot of chemical fertilizers. Will that not harm our land?"

The MP laughs out loud. "Aren't you a farmer? Don't you know that fertilizers are good for the land? They make the soil richer, and help you produce more crops. Even I know that." He turns to the people, who either laugh with him or shake their heads at me.

The chief says, "Ah, don't mind him, honorable MP. He is always like that, making trouble here and there. He even told my sons not to work in his old company because he said they will become sick. I think he is too lazy to work, but is also jealous of other people's success. I am thankful that my eldest son had the sense not to listen to him."

The MP turns back to me and says, "Well, I hope your grandmother will join this scheme then. It is an easy way for you to make a living." Spreading his arms wide as he faces the people, he says, "In fact, it is an easy way to earn money for all of you. You don't have to worry about your children and grandchildren anymore. They can work in your fields and stay with you."

Cheeks flush, eyes sparkle, and everyone speaks at the same time. The MP smiles like a benevolent uncle over his poor relatives as the officer tells them about the things they must do to prepare the land: chopping, clearing, and burning. Finally, he tells them what the government will provide them: saplings, technical training, and subsidized chemical fertilizers. The longhouse dwellers become so excited about the prospect of planting high-yielding palm trees on their land that barely anyone listens when the officer speaks about how palm oil is being used as biofuel in some European countries.

Finally, before the old grandfather clock strikes nine, the MP leaves with his entourage. The chief switches off the loud electric generator, which only makes the people's excited chatter more pronounced as they sit in circles of silent lamplights.

I walk through my grandmother's family room to the back of the house, burn a roll of fumigant bark, and step out into a sloping walkway just outside her kitchen. The symphony of singing insects and rustling leaves keeps me entertained as I gaze upon my moonlit surroundings. I love this time of the month, when the moon has waxed to its fullest and the people are compelled by the *adat* law to rest in honor of it. I love the moon so much; I think I must have visited the longhouse in the past months not to see my grandmother but to sit in her backyard to bask in the cold light. The air here is so fresh that my lungs feel light. I lean back, jealously watching a cloud that clusters at the fringes of the moon.

"Laja, what are you doing outside at this time of night? You are always like that. Everything you do brings nothing but trouble."

"It is such a lovely night, Grandma. You should come out here too."

"And get possessed by an incubus? I swear, you must have one of them in you. You never want to do the right thing."

I let her rant inside the walled kitchen, because I know that she will only stop when she is ready to stop. I feel restless. I sense that the forest echoes my uneasiness; or is it the other way around? I can never tell.

I will come again next month and the following month, and every month on the day of the full moon for the rest of my life. The graceful bamboos, feathery cassias, and fragrant grasses will still be here. I will still hear the long-call of the male ape as he announces his existence to man and nature. They will be here forever because they have been here since before I was born.

Nausea suddenly overcomes me. I retch over the side.

## CHAPTER THREE

# BEARING WITNESS

## Edged off Existence

Audrey Schulman, *Massachusetts*

On a beach along the Hudson Bay, I once snuck up too close to a polar bear. I only realized my danger when she rose to her feet and looked at me, thirty feet away. Polar bears are the largest land carnivore, weighing three times more than the average lion; if she charged me, there was little chance I'd make it to my truck. An animal like this could run down a galloping horse.

For several years before this trip, I'd been writing a novel about polar bears. The research had culminated in my taking a trip up to the Arctic to stand here openmouthed, facing a creature who on four feet stood almost as tall as I did on two. From my research, I knew that during the summer and fall there wasn't enough sea ice for the bears to cross the ocean to get to the ringed seals they ate. There aren't a lot of other animals around that could satisfy their hunger, so during the ice-free months the bears lost on average a third of their weight. It was October, and the animal standing in front of me was bony with starvation.

I'd felt safe enough to move in this close because she'd seemed so intent chewing on the body of a husky she'd dug out of the ground. From how flat and desiccated the dog's remains were—it looked like a rather stiff bath mat—it seemed clear that the body had been buried for quite a while. Unfortunately, I'd seriously misjudged the amount of time the husky would keep her occupied.

How can I describe the moment when she turned toward me with the same hungry intensity with which she had been chewing the dog? Unlike a shark, that swimming dinosaur, her reactions weren't hardwired. Looking into her dark eyes was more like returning the gaze of a gorilla or leopard, a creature with surprises. She watched me, mulling over the pros and cons of killing me.

Polar bears live on the edge, the edge of the world and the edge of their ability to survive even during good times. Spurred on by their harsh environment, they've evolved faster than almost any other mammal, diverging off from brown bears some 200,000 years ago. In this short time, they've perfected their arctic selves, becoming so well insulated that one of their main problems during the summer is staying cool. If you put on infrared goggles so your vision discerned only heat loss, when you looked at a human you'd see a silhouette filled in with psychedelic colors—the shirt and hair a neon purple, the open mouth yellow with heat, the aura of waste radiating out into the air. Furless humans are such immoderate folk. Now turn to look at a polar bear instead: the animal is utterly invisible except its nose, eyes and the slightly warm footprints it leaves behind.

The bears are superb swimmers and spend so much of their lives out on the ice that some scientists classify them as marine mammals, similar to dolphins or seals. They are capable of paddling eighty miles without stopping through the freezing ocean. Their paws—the size of dinner plates—paddle onward, their dark eyes scanning the water for an ice floe on which to rest.

They have a canny intelligence. Churchill, Manitoba, is the polar bear capital of the world, where the sea ice across the Hudson Bay forms the earliest each winter because of an unusual confluence of fresh and saline waters. In the fall, the bears wait there in increasing numbers for the ice to thicken enough for them to stride off toward the seals. Meanwhile, narrow with hunger, they nap in garages or stand outside bakeries sampling the air. Some of the worst encounters between humans and bears used to occur by the Churchill garbage dump, where a person stepping out of their car with their rustling garbage bags might find a polar bear padding out from behind an old fridge. In spite of their size and speed, though, the bears don't have rifles, so they were the ones who frequently suffered from any interaction. Decades ago, to help protect the bears, Greenpeace helped set up a garbage incinerator for Churchill so it could shut down the dump. Still, for three full generations the bears continued to visit the location of the old dump: the mothers taught their babies there had been food there, and although those babies never found anything there themselves, in this low-caloric world they taught their own children to continue to check out the possibility.

Three generations may also be all the time that the bears have left, according to current projections. The ice on the Arctic Sea has thinned more than 40% in the last five decades. By 2100, scientists predict that not enough ocean ice will remain for the bears to get to the seals. And this prediction is fairly optimistic. Given the way the Arctic is melting so much faster than we had expected only a few years ago, the bears will probably starve to death a lot sooner. Greenland's ice melt alone has doubled in the last five years. The ice floes on the open water are getting smaller and less frequent. Already, a winter survey of the bear population performed by the U.S. Minerals Management Service discovered the drowned bodies of bears bobbing in the sea. This was winter, in the Arctic, but the melting ice floes had drifted so far apart the bears couldn't find one to rest on, not even after swimming up to eighty miles.

These same surveys have shown that in the last fifteen years, the winter bear population has shifted from hardly ever being on land to being there almost 70% of the time. With the ice thinning, the animals are wandering far afield in desperate search of some new food source.

If you want to get a sense of how far polar bears live out on the margins of the possible—at the very edges of what a big carnivore can survive on—look at their pregnancies. The pregnancies are eight months long, and after that the mother dens herself up with the tiny pups for four to six months before she can hunt again. For those four to six months, the mother and babies survive on her fat reserves, never coming out to hunt. To decrease the risk that both the mother and cubs could starve to death in their den or that the mother might emerge too weak to hunt, the embryos don't implant in the uterine lining until halfway through the pregnancy. This way, if the pregnant bear hasn't managed to store enough fat to support herself and her cubs for half a year, a strategic miscarriage can save her life, to try again next winter.

According to NASA, over the last 25 years the average polar bear weight (both male and female) has decreased by 143 pounds, from what used to be the average weight of 800 pounds. We are literally erasing them from life, edging them pound by pound out of existence. Long before the last bear has starved to death or drowned out on the ice, there will simply be no new bears born for too many years in a row—all the young bears now middle-aged, moving past their fertility, every year all of the babies aborted in hopes of a better hunting year sometime in the future.

The bear staring at me that day on the shores of Hudson Bay was starving and desperate. Still, she sighed and turned away, letting me back away to my truck. She let me live.

Humans are not being half so kind to the bears.

## Ursus Maritimus Horribilis

Diane Gage, *California*

The ice cap is melting
and polar bears have begun to mate with grizzlies.
White offspring emerge

with black shadows under their eyes, long claws.
The bears look out
at too much sea for a polar sire and

too few trees for a grizzly sow.
They straddle the gap, adapt to a world with
diminishing ice and snow.

Hunters' helicopters whip the choking air,
burn Alaskas of oil.

On the other slope of the pole: more of the same.

## Thin Line Between

Marybeth Holleman, *Alaska*

*for Tom*

In the slant light of late afternoon, the polar bear's fur is not white as snow but a silhouette of buttery tones against bands of pale blue sky, granite sea, and marbled sand. Wet from her swim across the channel, her fur spikes along the curve of her back and ripples as she pads the shoreline.

She mouths a scrap of whale skin, flipping it into the air, then sniffs another piece and pushes it over with one swipe of her great paw. Ten feet away, from inside our truck, we watch, silent except for the click of cameras, until she walks out into shallow water, hopping lightly among a drift of gulls that scatter before her.

This, my first sighting of a polar bear in the wild, comes less than an hour after my son, husband, and I arrived on Barter Island at Kaktovik, an Inupiat village wedged between the Arctic Coastal Plain and Beaufort Sea. Six years ago, in 2003, when I first heard about and decided that I wanted to see these polar bears, they were considered to be a healthy population, safe on their sea ice. Two years later, I heard the first reports of polar bears drowning as they tried to reach a rapidly shrinking summer ice sheet. Since then, the news has spiraled much more quickly than I could have imagined, so that this trip is now shot through with sorrow: according to current projections, Alaska's polar bears will be gone within thirty years.

We've arrived near the end of the bone-pile feeding season. Kaktovik whaling crews set out in their *umiaks* just after Labor Day, and usually reach their quota of three bowheads within weeks. After cutting and distributing the whale blubber, *muktuk,* among the 200 Inupiat residents, they load the whale bones, plus a fair amount of skin, sinew, and gut, into a pile at the end of a three-mile-long sand spit that arcs out into the sea in front of the village.

That's what brings in the polar bears. The spit fans out at the end, a swirl of sand littered with the bone piles of previous years. These old bones are weathered to a pale grey patina, in sharp contrast to the new pile, blood-red and ragged. Together, they form a Dali-like sculpture, a monument to the millennia-old relationship between Eskimos, whales, and polar bears.

The bears arrive earlier every year, eager to scavenge whale carcasses, restless from their long summer hunger. Even before the umiaks are launched, their white heads emerge from the water, scanning the shoreline and sniffing the air for the pungent scent of blubber. This year, the whales are much smaller, and so the fall feeding ends earlier. Villagers prefer small whales because they are tastier and easier to handle, but smaller whales mean less for the polar bears. During our four-day stay, the numbers of polar bears on the bone pile dwindles down to one, and then, on our last morning, none.

I feel no small amount of ambivalence about this trip, so much so that I almost cancelled it. It seemed hypocritical to fly the 650 miles from Anchorage to Kaktovik just to see the animals who are dying because of the very fossil fuel expended to bring us here. It seemed like I was making a choice, by getting on that plane, that I was personally signing their execution papers.

Lurking beneath the surface, there may have been another reason to avoid going to see the bears: maybe I didn't want to look them in the eye, knowing what I know. Maybe I'd rather let it all remain abstract, polar bears and shrinking ice packs. Maybe I wasn't ready to face the truth.

I had wanted to see them for so long, and that desire had grown more desperate with the bad news. Perhaps it was selfish; perhaps I wanted to check them off my life list while they were still around. Still, I also believed that, by seeing them first-hand, I might have some insight about how to save them, insight that would, in the end, be more useful than simply not getting on the plane.

Watching that lone female polar bear meander the shoreline on our first afternoon here does not, as I had secretly hoped, vanquish my ambivalence. I feel instead as if I am leaning into the sharp point of reality itself.

In the morning, my husband, son, and I drive through the village to the back side of Barter Island. Beside many of the houses, on wooden platforms or blue tarps, are rows of muktuk cut into large squares. Already, in late September, temperatures have dropped enough to use this natural outdoor freezer. One slab of this whale blubber lies in the middle of the road, fifty feet from the nearest outdoor stash. We learn that a polar bear entered town in the night and snatched this piece, but the Village Public Safety Officer scared it off, and the muktuk still lies where the bear dropped it.

At this time of year, when the polar bears gather at the bone pile, the VPSO patrols the streets every night for brazen bears. "They're looking for a last meal before winter hibernation," says one resident. "But if they're not careful, it will be their last meal, period."

I can't help but wish that the polar bear had gotten its muktuk. I wonder what is to become of these hungry bears now that the whale carcass feed is over and the sea ice is still nowhere in sight. On our way out to Kaktovik, we ran into a friend who does fly-over whale surveys for the federal government. She said the pack ice was still over 250 miles from the coast. She saw a couple of bears swimming from shore, and wanted to call down to them, "Go back! It's still too far away!"

Dusk falls as we head again to the bone pile. This time, we're in a small school bus with two other visitors and a driver, Art. It's cold, perhaps twenty degrees, and a stiff wind whips across the sands. We quickly count seven polar bears on and around the whale bones. On the far edge of the spit, another bear climbs out of the sea, shakes off water, and heads for the bones.

The bears are surprisingly tolerant of each other, until one finds something at the edge of the pile. Another bear notices and approaches the prize with its long neck lowered. The first bear snarls, shakes its head, and slouches away. The other bear tries a few more times and is met with growling, paws raised, teeth bared in warning. Then it's over.

With the engine cut, we're so close that we can hear polar bears tearing bone apart to get at the marrow. No easy pickings of blubber or meat remain. Some of the bears, their snouts stained henna red, are half-hidden behind jaw and skull and rib bones. It's a primal scene: the windy cold, the piles of bronze-colored bones slick with whale oil, big bears crunching marrow, and the musky, acrid odor of dead whale. Behind it all, a slice of moonlight glistening off waves and the thick line of an orange sunset intersect.

Suddenly, one bear is right beside us, walking around the bus. He puts his front paws up on the hood and then drops down, saunters around to the side, rises upon hind legs and appears at the window, where I stand. We all jolt from our collective trance to realize that the windows are down, and the bear's head is through the opening, those small dark eyes, black wet nose, white face, right there. Inches away. With his ears pressed back and his eyes wide he looks so—well, submissive. I check an impulse to reach out and stroke his head.

Art starts the engine, yells "Bad bear!" and honks the horn. The bear staggers from the window, drops to all fours, and backs away. He tries once more, jumping up and placing both paws on now-closed windows, fogging the glass with his breath. Then he drops down and saunters toward the water.

We are jittery, joy and fear so intermingled that they're indistinguishable. The bear has left a perfect paw print in the dust on a back window. I lean over and look through it. There, in the sky, are shimmering ribbons of red and green: the northern lights.

Art says that bear is way too bold, and will most likely get into trouble in the village if it doesn't watch out. This is bound to happen more often, though: polar bears entering human settlements in search of food. The most famous place to see polar bears, Churchill, exists not only because it's one of the first places where shore ice forms but also because of the town dump. With shrinking sea ice, with more polar bears stranded on land, these kinds of encounters will rise.

It already seems inevitable, irreversible. This is the thing about global warming that knocks the breath from me: as soon as I learned the depth and breadth of it, it was too late. Just a decade ago, it loomed on my peripheral vision as no more than a vague theory. Now, suddenly, it's too late—at least for polar bears, and a host of other species and habitats.

Every other time I've learned of an animal in trouble, there's been enough space between *threatened* and *extinct* in which actions can save them. Protect habitat, and black-footed ferrets survive. Ban DDT, and peregrine falcons return. With global warming, there is no such space between.

I came of age at the time of Chernobyl and Three Mile Island. In my childhood, I saw black bears chained outside convenience stores for the pleasure of passing tourists; I glimpsed cougars disappearing into thick mountain fog. On the first Earth Day, I collected litter in my neighborhood and sealed my fate. In high school, I witnessed the forests on Mount Mitchell die from acid rain. In college, I joined marches on Washington, D.C., against nuclear power, and rallied against a dam that would flood a wild bottomland in North Carolina. In Alaska, I volunteered to rescue oiled birds and clean the beaches of Prince William Sound; I spoke out against aerial gunning of wolves and bears, against clearcut logging and large-scale mining.

Global warming blindsided me. I've spent my whole life tilting at windmills, while a thunderous cloud gathered overhead, and I simply failed to look up and see it. The polar bears are still here—I sit in a truck and watch them feed on whale bones a few feet away—but they are disappearing, and I didn't even get a chance to fight for them. My dawn of recognition came precisely at nightfall.

On our last night at the bone pile, an Inupiat man on a four-wheeler appears and circles us, then stops and asks Art, "Are you stuck?"

"No," says Art. "We're waiting for the bears."

"Oh," the man says, and turns on a large spotlight and scans the shoreline. He clicks it off, turns back to us, and laughs, "Where'd you put my bears?"

He scans the sea with his light again, and points out the reflective eyes of one bear, swimming toward us in the dark sea.

Kaktovik residents tell us that they have a different relationship with polar bears than do their neighbors in Barrow and villages further west. They don't kill them for nearing the village, though the law would allow it. They have chosen not only to tolerate but to enjoy their polar bears, and to share them with a handful of visitors.

When I ask why, I'm told that a couple of elders decided that they liked watching the bears on the bone piles and didn't want them killed. The rest of the village followed their lead, and now it's a habit. In spite of these bears roaming their village at night, they still maintain equanimity toward them, a kind of gentle respect. We often saw one or two other vehicles at the bone pile, residents spending the evening watching their bears.

By contrast, in Barrow, the regional population center for the North Slope, a polar bear roaming near town is a dead bear. With their larger population, Barrow whalers are allocated fourteen whale strikes, but to keep bears away from town, they bury the carcasses. I wonder what would happen if instead they moved the bones farther from town. I wonder if bowhead remains piled up and down this coastline could serve as a stopgap measure for the polar bears, giving pregnant sows a good last feed before hibernation, giving the others something to eat while they wait for the ice, which will take longer every year.

I'm grasping for solutions, frantic in my desire to do something to stave off polar bear demise until we can slow and reverse global warming. That's why I'm drawn to the tolerance that Kaktovik residents show their bears. I'm searching every story for clues to salvation.

One Native man we meet at a church service, a tribal elder, tells us about an Inuit tribe in Canada that had a stash of hundreds of eider eggs, set aside for winter. One night, a polar bear discovered the stash and ate every egg. As he tells us the story, though, the elder shows not a hint of anger at the bear for taking the Inuits' winter food. Instead, he just laughs and says, "Happy bear!"

In the ensuing months, my worst fears continue to be realized. Within a week of each other, two events: A solitary polar bear enters the village of Fort Yukon, 250 miles from the coast, the farthest inland ever seen in Alaska, and is shot by two residents. In Canada's Northwest Territories, a sow and two cubs travel 300 miles inland to the southern shore of Great Bear Lake. They charge some town dogs in attempts to steal their dog food, and the bears are shot dead by police.

Are these my bears? The ones I met in Kaktovik? Are they the sow and cubs I watched on a barrier island? The lanky male who tried to steal a scrap at the bone pile?

When I learn that scientists predict that two-thirds of the world's polar bear population will die off in the next couple of decades, I wonder if that includes the female that I saw walking in the shallow waters off the spit. When I'm told that this year's cubs may well be the last born in Alaska, I think of the bear who stuck his head in the bus window.

Seeing polar bears hasn't assuaged my fears for them or given me any new insights that might save them. It's made them more familiar, more difficult to lose. I hear bad news, and I don't think of some homogenous group called "population." I think of the bears I met.

From our brief time in Kaktovik, the polar bear images I treasure the most aren't the close-up encounters at the bone pile but the long view through binoculars of all the polar bears, together. The coast at Kaktovik is a winding waterway between barrier islands, curves of sand and mounds of grasses and willows. During whaling season, when they're not on the bone pile, the bears assemble on one nearby sand island. There were days we counted up to twenty-four at a time.

A sow with two cubs walked toward the gathering, slowly, as the cubs repeatedly stopped to play, one rolling over on its back like a dog, paws in air, and the other climbing up, tumbling off. Two adolescent bears wrestled, one climbing onto a large piece of driftwood and pouncing onto the one below, toppling over in slow motion with that great soft body. Another sow with cubs slept curled together into one big white hummock.

This congregation doesn't fit with what we think we know of polar bears; we usually describe them as solitary animals, only tolerating each other to mate and then to raise young. Perhaps this is an aberration brought on by shrinking ice pack and diminished food sources. Or perhaps they've always done this. What do *we* know?

What do *I* know?

I no longer hold some Pollyannish belief that polar bears will somehow survive, that next winter the pack ice will miraculously grow, that the flat-Earth crowd claiming global warming is a myth may be right. Neither is my sense of loss alleviated when I learn that a small percentage of the polar bear population may survive in some corner of northeastern Canada. That doesn't help *my* bears.

This I do know: life wants to endure. Polar bears traveling hundreds of miles inland, far from their sea, looking for food; polar bears roaming the village at night, trying to steal a piece of muktuk: they are expressing their will to endure.

The world is not and never has been fair or just or right. It just is. Until it's not. Polar bears may go, too, and it doesn't matter what I say or do, but I won't give up. Because life wants to endure. *Because this is what I want to do with my life.* Seeing polar bears, watching the moments of their lives, reminds me that's all there is: this moment. Now. And now.

Life was never a given, the future was never assured. Each day really is a gift, one given freely with no strings attached on either side: no requirements, and no promises.

There's not much in this world that is always true. But there's that.

## Polar Bears

J.R. Solonche, *New York*

Dying off. That's what's happening.
They're dying off. Is this really happening?
Their kingdom of ice, their white continent,
their fortress, their palace, their world
is melting, dissolving around them,
disappearing out from under them,
and they are drowning in it, the solitary males
and the females with their cubs, they are all
drowning in it, this desert of water whose
distances now are too far for them to swim
so they tire and drown, this iceless sea
with no solidity that we have made of their
kingdom of ice, not thick enough, not deep
enough, not strong enough to save them
from us. I cannot believe
what we are doing to them, what
we will have done to them when it is done,
to the polar bears, to the animals I could
not get enough of, in the Bronx Zoological
Garden, when my mother took me there
to see them, as I stood with my hands
on the iron railing and cried out, "Look
at the white bears! They look like snow!"
and stood, awestruck, with my hands stuck
to the iron railing, just stood there and looked
at them, at the white bears, at the bears
that looked like snow, and looked at them,
and looked at them, and looked, and looked.

# Part II
# Generations

*As in the last poem of the preceding chapter, for many of us, some of our deepest experiences of the natural world have come in childhood and youth. But what is it like for children today to grow up on a planet whose very foundations are being shaken? What does global warming mean for our children's senses of innocence and of strength, their capacities for wonder and for hope, their trust in their parents and in themselves? What world will we all inherit in the coming decades and centuries—or maybe as soon as next year?*

*After the observations of the present day offered in Part I, Part II explores the effects of global warming on future generations. Chapter Four, "The Gifts We're Giving Our Children," considers some of the common experiences and quandaries with which parents and children are struggling, together and alone, as we all face the losses and tragedies of a changing planet. The chapter's selections proceed from younger children to older, all of whose lives today are marked and scarred by the psychic burden of an uncertain future. Shifting to the lens of fiction, Chapter Five, "Future Imperfect," looks farther ahead, imagining loss, guilt, desperation, and survival in a world made increasingly chaotic and dangerous by climate change.*

CHAPTER FOUR

# The Gifts We're Giving Our Children

## Learning Their Names as They Go

Kristin Berger, *Oregon*

An inverted *V* below the nose is the sign
we use to say *Walrus*—
two tusks beneath scruffy cheeks,
like an old man with kind eyes
waiting to be noticed. He is
one more illustration in a child's book
full of disappearing wonders—
*beluga, narwhal, murre*—
their spoken names the receding mantras
we also learn, more words than species,
more ways of saying the thing that is
becoming less than paper and ink.

But my daughter loves him, the Walrus.
How could she not,
floating on his thin berg
towards the hot, open sea?
Love at first sight and every glance thereafter—
she points him out on each page,
his toothy simplicity mirroring
her mouth still accumulating pearls,
her emerging, buoyant wonder:
She makes the sign
that is more-than-Walrus,
more than even his own name
for his melting, heavy-hearted self.

*This poem was written in the aftermath of Hurricane Katrina.*

## The Darkness

Lilace Mellin Guignard, *Pennsylvania*

"I can make the dark not dark anymore."
His shoes blink red lights against the pavement.
My son's hand in mine, his night steps are sure.

Such confidence. Such trust. Just what's in store
for those unaware their future's been spent?
"I can make the dark not dark anymore."

He thinks it's easy. Tell that to the poor
folks un-homed by hurricane and President.
My son's hand in mine, his night steps are sure.

It's hard to keep my faith strong at the core
when I know how bad the earth has been bent
to make the dark not dark anymore,

the hot more cold and cold more hot. Ignored
consequences creep up with a vengeance.
My son's hand in mine, his night steps are sure

that I'm in charge. How to prepare him for
how progress turned to global accident?
I can't make the dark not dark anymore,
my son. Hand in mine, your night steps sure.

## The Innocence of Ice

Jamie Sweitzer Brandstadter, *Pennsylvania*

Spirals of smoke rose up to meet the clouds each time we exhaled, and the sunlight glistened in the beads of sweat that our bodies had expelled despite the cold temperature. There was laughter; there was the crack of a stick as it propelled its small, round victim; there was sometimes blood, but rarely tears; there was always the over-arching dome of innocence, isolation, and fulfillment that came from a winter's day on our grandmother's pond.

The ritual had survived for generations. The swish of the ice slowly chipping around metal must have been addicting, for the game had claimed its victory over each new electronic device—from Atari to Nintendo—that threatened to distract each new group of children from matching our blades to the top of the frozen water. We were kids playing hockey on a pond; we had makeshift goals, hammered-together pieces of plywood or soup boxes that were found in the third floor of the barn; we had hand-me-down skates that had been laced and re-laced since the early 1950s; we had hand-me-down sticks that were held together by electrical tape and twine; we had Carhartt jackets and retro earmuffs. But we were champions.

The rest of the world walked off stage as the first blade hit the ice. The games would start early in the morning when the full sun finally set the ice on fire; they would end long after it had been extinguished by the hill littered with countless decaying and collapsed stalks of corn. Maybe it's ironic, or worse, that many of the best games were played by the glow of the dusk-to-dawn light whose consistent, unconditional love was powered by coal birthed in that same Pennsylvania valley, or the valley over. On other nights, all we needed was the light of the stars that barely outnumbered the memories of generations of plow-driving farmers by spring and body-checking hockey stars by winter.

Teeth were lost on that ice—a small body crumbled to the

edge, digging hopelessly through the pristine white snow for his treasured incisor—but not much else was lost. The rest of the planet seems all too familiar with and capable of loss, but somehow, on that ice, all was salvaged, as if shielded in a dome of innocent preservation. On the ice, time itself seemed suspended, as seven-year-olds became aggressive, agile left-wingers, and forty-somethings became children once again.

Today, though, the innocence of the ice has not been preserved; time and a change never known before have depleted this archaic wonder. It has been years since a skate has touched the ice or a slap shot has hammered into the back of the plywood goals. College, work, or distance have not ended this family tradition, but something of a more menacing nature has—an unfamiliar, unnatural force. We have melted the ice. We have forfeited our own hockey games. The rest of the world has stood up, stolen the limelight, and thrown its arms up in a cruel victory against what once was.

I dread the drive into the wintry valley, with little snow and a half-submerged makeshift goal. I dread the sight of dusty skates hanging on the cement basement wall. I dread the sense that we are forgetting what we once shared, and we are becoming victims of experience. I dread the fact that all I will have to tell my sisters' children, my children, will be tales of a frozen hockey rink where championships were won, where time stood still, and where love existed solely.

## Annapolis Bus Ride

Julie Dunlap, *Maryland*

Eli inched closer to me on the bus seat as the two young men behind us grew more agitated.

"I can't believe how bad the news is getting!"

"Yeah. And politicians do nothing while the planet goes to hell."

I distracted my eight-year-old from listening to more of their conversation by pointing out sights through our window—the Treaty of Paris Inn, Prince George Street, the Maryland State House. A home-schooled history buff, Eli was obsessed with the Revolutionary War. I recounted a few hazy facts about where George Washington resigned from the Continental Army and why the port town was a temporary capital of the new nation. As Eli gazed silently, absorbed, I imagined that he saw lads in three-cornered hats riding beside us, all on our way to the climate change rally.

My mind wandered to a more recent past, to the bus ride through Illinois that ecologist Aldo Leopold took in 1944. While farmers on board chatted about baseball and taxes, Leopold mourned the decimated prairie grasses, the absent quail, and worst of all, his fellow passengers' ignorance of all they had lost. "One of the penalties of an ecological education," wrote Leopold, "is that one lives alone in a world of wounds."

Connecting with others was one reason I took my kids to climate protests and clean energy rallies like this one. The bus discharged us, along with our expressive young compatriots, onto a plaza teeming with sign-toting, petition-passing activists. We were welcomed with stickers demanding *Warming Solutions—Now!* and green hard hats that symbolized support for a sustainable jobs bill. An organizer spotted Eli and hustled him in front of the podium; a smiling redheaded boy might charm a television camera crew to focus on the poster he was waving. I grinned with pride as any stage-mom would.

Local luminaries took turns at the microphone, each dispensing at least a modicum of alarums—rising temperatures, eroding shorelines, drought and hurricanes, extinction, hunger, and political unrest. Eli listened politely, showing no signs of distress. Though only in the third grade, he had heard it all before. Television, web sites, science lessons, and even nature centers bombard young children with global warming scenarios. That week's issue of Eli's favorite magazine, *Time for Kids*, followed a dogsled expedition to collect data on the thinning Arctic ice cap, informing readers who still believe in Santa that "the North Pole could be underwater during the summer in less than 10 years." Reading over Eli's shoulder, I had agreed with musher and activist Will Steger that children's awareness should be raised to spark action against greenhouse gases.

The next speaker opened with a joke. "The good news is—we have beautiful, warm March weather for our rally. The bad news is—we have warm weather for our rally in *March*." Eli chuckled along with the crowd, but I shuddered. It hit me then, as it had for different reasons on the bus, that perhaps this rally was not the place for my child. By his ninth spring, my son had learned to doubt the reliability of the seasons. The thought staggered me. Rachel Carson once wrote, "There is something infinitely healing in these repeated refrains of nature—the assurance that dawn comes before night, and spring after winter." Carson drew great strength from migrating birds, budding flowers, and other timeless rhythms of spring, but we are rearing a generation to expect little from the earth but uncertainty.

Part of the answer must lie in expanding environmental education beyond words and images on page or screen by reconnecting children with the living world outdoors. In his book *Last Child in the Woods*, Richard Louv warns that many children spend more time with Nintendos than in their backyards, and kids raised indoors pay a price that can include obesity, depression, attention deficits, and waning spirituality. His prescription to cure "nature deficit disorder" is active outdoor play, anything from hiking and canoeing to bug collecting and tree house building. "Get kids directly engaged in nature," says Louv. "Hands dirty, feet wet." Indeed, I cannot doubt the healing effects of sky, trees, and pavement-free space whenever I take Eli and his sister to our favorite park. As a homeschooling parent, I can grant them the time to overturn rocks in a stream and poke sticks into tree holes. Such activities nurture human connections as well; outings with friends and families who share our attachment to unmown places build community, perhaps better than chanting slogans in a crowd.

Lately, though, when I sign the kids up for guided nature walks, I risk the leader pointing out scars and diminishments in the landscape. "See this vine with porcelain-blue berries?" one asked, drawing the kids in closer, their eyes widening in anticipation. "It's an invader that chokes native species. More plants like that will take over as the climate changes." The children stepped back. Like an apocalyptic rally speaker, this gloom-focused leader may, bit by bit, have undermined Eli and Sarah's love of nature and faith in their future. I harmed them, too, by not whisking them back to the stream. Preserving a capacity to wonder must be as vital to conservation as protecting the land itself. My children, thanks to such acts of commission and omission, are growing up in a world of wounds.

Standing at this rally, I watched activists shout and chant and bolster each other toward needed change; for them, action is a balm to anxiety. Yet for vulnerable young minds, exposed daily to disturbing reports and terrifying predictions, we need to find new ways to protect and to heal. It's not that we don't know or care about our children's inner sensitivity in other arenas; a few curse words in Eli's vicinity motivated me to quiet intervention, and most parents I know would stop kids his age from seeing violent or sexually explicit movies. Responsible adults routinely shelter pre-teens, if not older children, from the Holocaust, the Rwandan genocide, and other examples of humanity's darkest crimes. So why don't we safeguard our youth from this ominous information about nature itself?

More stressful even than age-inappropriate messages are calls for children to take responsibility for resolving the crisis. "Will Steger and his teammates," says *Time for Kids* about the polar expedition, "want to jolt kids into action on global warming." An internet contest for children asks, in cartoon graphics, "Got an idea that will save the Earth?" My thirteen-year-old shook her head ruefully and went back to re-reading *Harry Potter*. Aldo Leopold contended that "an ecological conscience . . . reflects a conviction of individual responsibility for the health of the land." But he did not intend for the youngest generation to carry the burden for their elders.

Perhaps child development experts could arbitrate when to introduce climate change issues, and how soon to expect kids to get involved. Apparently, researchers have not yet begun to investigate the question. But surely kindergarteners do not need to know that human effluents are reducing the habitability of their planet. Children who are not yet able to think abstractly about science, politics, or economics should not be asked to invent ways to restore the atmosphere. Even when global warming lessons are developmentally appropriate, schools should present them with as much preparation, monitoring, and parental oversight as middle school courses discussing contraception and STDs.

Of course, the ubiquity of the climate change message makes it tough to avoid. To limit all sorts of frightening visuals, commercial propaganda, and political misinformation, author Barbara Kingsolver bans television from her home. Despite her daughters' objections, Kingsolver accuses the "one-eyed monster" of stealing time and distorting reality, especially TV news, for it "provides a peculiarly unbalanced diet for the human psyche." Instead, her family listens to the radio and checks farm forecasts and obituaries in the *Country News*. They find baking casseroles for bereaved neighbors a more responsive act than railing against intractable national policies.

Certainly Kingsolver never positions her girls center stage at environmental protests. Nor can I imagine her souring their affection for the Kentucky hills with lectures on the extirpated American chestnut. But Kingsolver offers no advice on what age group might be mature enough to stop protecting. Scott Russell Sanders has written about his seventeen-year-old son's anger and despair over the state of the world. "The first condition of hope," says Sanders, "is to believe that you will *have* a future; the second is to believe that there will be a decent world in which to live it." Though groomed to ace their SATs, even the eldest children may not have learned to hope.

Aldo Leopold lamented seventy years ago that conservation education had failed to convince farmers to cherish their heritage of oaks and prairie flowers. He faulted not the quantity of effort but its content, calling for attitude changes rather than more facts. But many twenty-first century kids have learned to care about the earth, some too much for their own emotional well-being. These days, it seems we need educational content that instructs children in hope—including day-to-day outdoor experiences that offer it naturally.

The lessons may differ for each child. Studying geology or the stars might give some a deeper perspective. For Eli, I will look toward the past. Instead of staying on the rally bus, I could have stepped off to show Eli the Annapolis home of William Paca, who risked death as a traitor by signing the Declaration of Independence. Such examples can reveal not only individual courage in a volatile world but also an uplifting societal response to a shared crisis. As Abigail Adams said, "Great necessities call out great virtues." We can tell children that humans have confronted challenges before and mustered the collective will to prevail.

I will also take Eli further back, to a riverbank, wetland, or the wooded edge of our backyard to learn hope from what Leopold called the "odyssey of evolution." Beyond outdoor play, children may need to pursue serious study of life's history before they can appreciate its intricacy and understand its resilience. To experience that deeper history, Leopold abandoned country roads for the marsh. "A sense of time lies thick and heavy on such a place," Leopold wrote. "Yearly since the Ice Age, it has awakened each spring to the clangor of cranes." In hearing the croaking cries of sandhills on the marsh, Leopold regained a deep-time perspective of nature's perseverance and enduring seasons our children need to thrive. These living remnants of the Eocene were "the symbol of our untamable past, of that incredible sweep of millennia which underlies and conditions the daily affairs of birds and men."

The farmers on Leopold's bus ride neither worried about yesterday nor imagined a finer tomorrow. They were oblivious, said Leopold, for to them "Illinois has no genesis, no history, no shoals or deeps, no tides of life and death." The young protesters on my rally shuttle, not too far from childhood themselves, may have felt guilty about our collective past, and had gathered to plead for a future—but they didn't seem to believe in one. Like other parents, I want more for the children I raise. Though I cannot protect them from every wound, I can help them heal. Better still, I can help them trust that health remains possible. It may be a few years before I take Eli on a protest bus again, and there are no bogs nearby graced with the voices of cranes. But we can head to a salt marsh and linger for hours, watching the endless rise and fall of the tides.

## The Last Days

Dane Cervine, *California*

My daughter looks up from the Sunday news—
an earthquake in Pakistan, the many dead—
betrays a quick glance of fear, after so many
hurricanes these last days, New Orleans flooded,
Texas evacuated, Florida bracing, Indonesia still
reeling from the last tsunami. The book
of Revelations lies in my childhood memory,
prophecy of flood, famine, fear—but I
can't bear to tell her my secret misgivings,
that I am nearly fifty, peering down the gauntlet
of my own last days, wondering how to spend
judiciously, extravagantly, each one of them.
But this is all so personal, a sin, really,
when living in the belly of an empire
bent on catapulting us into the next war,
the next bald-faced robbery of a planet's future
for this year's money-grab, the whole world
aghast and envious of this drunken bully
staggering belligerently towards oil
like an addict who would do
    anything, anything
for just one more fix. But I am no prophet,
succumb to the small world of breakfast
my daughter and I share, intimate,
in these last days of childhood, poised
as she is on the lip of a world that would
as soon devour as kiss her, and
how can I prepare her for this rogue?
The way fear's scripture insinuates its way
into the petty concerns of a life lashed
to the mundane but longing for revelation,

for some final reckoning. How do I say
*I love you* when this world is the only gift,
the pitiless dowry, I have to offer? How
do I say *it is you*, your brother, your friends, that are
the only hope for this brooding planet,
a new seed of reluctant messiahs peering
at the earth you shall inherit from us—
this thin, ephemeral line between
Eden and Armageddon?

CHAPTER FIVE

# Future Imperfect

## The Last Snow in Abilene

Benjamin Morris, *Mississippi*

In the end, it ended
without a fight—

        no radio
squelching the news, no TV debate
or learned expert explaining

the fact until years later,
when the frost, too, failed to come.

That spring an army of daffodils
besieged the courthouse, and the cats
slept outside in abandon.

At some point,
no one remembers now,

        they understood,
and schoolchildren began to ask
*who sent it?* and *why did no one save some?*

The old-timers in town,
clustered like horses
round the watering hole,

shook their heads
and reminisced about the last time
they had seen snow

          crowning the hills:
thick plumes so clean and cold
they stilled the wind, snow

that, when drizzled over mint
and a little whiskey, held to the light,

would glint there, and glisten,
and melt into the night.

*The following is a haibun, a genre that combines prose and poetry (haiku) in expressive rather than strictly narrative ways, evoking widening ripples of association. Originally from Japan, the genre now has practitioners worldwide. (Indeed, one might think of this whole book as a sort of haibun.)*

## Blue Sky

Penny Harter, *New Jersey*

On weekends, she walks up hills to see the sun. At sea level, thick smog obliterates the sky, a gray and toxic smothering. Despite the altitude, once she gets above the gray she breathes easier. She has not seen such a blue sky from down below since childhood.

masquerade party—
strangers crowding into
a downtown loft

When she tries to get some of her co-workers from the factory to climb with her, they merely laugh. "But you can see the sun," she exclaims. "And the sky is blue!" Her friends prefer the mall or the movies, so she climbs alone.

shooting star—
how briefly its wake
marks the dark

Years pass, and she has to climb higher and higher. Having retired, she can climb more often, but it's slower going now. One day when she arrives above the timberline, stumbling among rocks shining with lichen, she is breathing in stabbing gasps. Soon she will be too old for this, she thinks. Head spinning, she clings to a nearby boulder and stares up into the blazing heavens. She looks down at the tide of gray creeping up the slopes. She knows it is only a question of time until she will be forced to go up and up.

moon colony—
again, the supply ship
arrives late

## After

Jo Salas, *New York*

You want me to tell you what it was like, before? Of course I remember. Sometimes I wish I didn't, that my old brain had dried out after all these years.

Well then. I remember the summers. How we welcomed the warmth after the cold winter. I remember having dinner outside almost every night, on the deck. You could eat like royalty from local farms—corn picked an hour ago, bursting with sweetness, voluptuous tomatoes, beets and beet greens, blueberries and peaches. Sometimes we had more than we could eat, can you imagine that? I would chop up a big bundle of fresh Swiss chard and throw it into the boiling pasta water a minute before the pasta was done. Then toss it all together with gorgonzola and sautéed walnuts.

What's gorgonzola? Oh—something we used to eat. A kind of cheese. It was particularly delicious. Tangy and soft. It came from Italy. Well, you asked.

Sometimes a thunderstorm would roll in from the hills but we'd stay outside, under the deck's roof, relishing the clean wind and the cooling raindrops on our backs. The air was so hot. Each year hotter for longer, it seemed. We got special shades to insulate the windows, then air conditioners, when the shades were no longer enough.

The cats would stretch contented beside us as we ate and drank wine and talked, but when they heard the thunder they'd run inside to hide. Safe as they felt with us, nestling bonelessly on our laps, their little cat brains told them that we could not protect them from the dangers of rumbles in the sky. I felt so tender towards them, those two cats. Ben and I used to tease each other that they were like our children. We were sure we'd have real children one day. We were young, not much older than you all are now. We thought we had time.

No. Don't ask me. I can't think about that.

Come a little closer, please. My voice gets tired. No, that's close enough.

In those long evenings, the flowers on the deck were luminous and fragrant: white and purple petunias, nasturtiums, phlox, alyssum. Pansies. The morning glories rolled themselves up tight for the night, waiting for the morning sun when they'd open out again into their triumphant blue. I planted masses of flowers on the deck each spring and they'd bloom all the way into October. There were deer, you see, lovely to watch but they ate everything I tried to grow in the yard. Eventually I gave up and turned the deck into my flower garden, and surrendered the yard to the deer. They seemed to assume the apple and pear trees were for their sole benefit. We didn't mind. We enjoyed watching the deer families gather at the end of the day to eat fruit that had fallen to the ground, as well as whatever they could reach by craning their necks up. Sometimes an enterprising young one would stand on her hind legs to reach a tempting cluster.

No, we didn't want to kill them. We had enough to eat. We were never hungry.

There were so many birds, too, finches and sparrows, cardinals, even bluebirds. A hummingbird who'd visit each flower on the deck, its invisible wings whirring so fast you could feel the vibration. Sometimes a flock of wild turkeys strutting by. Hawks cruising on the air currents along the cliffs. The birds were a constant presence around us with their singing and their flight. A blessing.

Well, it may sound like paradise to you. In a way it was. We just didn't realize it. There were lots of things wrong in our world, no shortage of misery, but there was paradise too. And it's gone.

Our house was on a country road not far from here. A small ordinary house with a rambling yard. There were other houses around. Empty now, of course, and half ruined.

Yes, we knew our neighbors. We were friends with all of them, except the man who insisted on spraying chemicals on his lawn. So many people were ill, even then, before all the pandemics. I feared for Ben more than myself, though he seemed perfectly well. A foreboding. I hated the idea of carcinogens drifting in our windows and seeping into the water table and our well. Not to mention poisoning the wild creatures and the ground itself. The neighbor said I had no right to comment on how he kept his lawn. He harrumphed about property rights. I suppose now, if he's not dead, he might agree that I had a point.

They lived in the city, that couple, and drove up every weekend, about eighty miles each way. They had an enormous car that must have used a great deal of gas. But almost all of us were guilty in that regard. We had to drive our cars everywhere because, except in the cities, there was no other way to get around. If I wanted to go into town I had to drive.

Please don't shout at me. We didn't think we had a choice, in those days. I know it's hard to believe now. A few people tried, but not enough. Don't you think we're paying the price? In some ways it's worse for us than for you young ones. At least you're not shattered by guilt. You're not tormented by remembering.

Where was I?

One thing that's still the same is the cliffs. Sometimes I walk to the edge and look down and I think about how ancient they are. They've seen animals evolve and they saw the first humans to come here, and they see what we've come to now. They'll outlast us, too. I could see them from my house, in the distance. I used to watch from my bedroom when the sun first rose, waiting for that brief spell of pink when it looked to me as though they were radiating light rather than reflecting it. I would watch, and then when it was over I'd get up for my shower. I loved the flow of hot water on my skin. It was a ritual. Sometimes I'd let myself stay in the shower for a long time, for the sheer pleasure of it. We all used as much water as we wanted, for washing, for drinking, for plumbing. I remember reading about how much water we used compared to someone in a poor country, and I was shocked, but I didn't see how I could help. I could hardly send water to them, could I?

You're right. We did take a great deal for granted. We didn't think it would ever change. We simply couldn't believe it would change, though we were warned for years.

Ben used to leave early for work. I was alone for breakfast, and I usually had the same thing. Brown bread, toasted, with butter and jam, and a cup of Irish tea. That's all. Perfect.

I'm not crying. Leave me alone. Do you think I'd cry about breakfast?

Well. Sometimes I had music playing while I ate my breakfast, a recording of Bach or folk songs or music from Africa, whatever I felt like. It was simply a matter of pressing a button and it sounded as though the musicians were right there in the room. Or I'd watch the news, though it got worse and worse and finally I really couldn't bear it. So many terrible things that I couldn't do anything about. The music was much better than the news. The music always made sense. All day long, if I wanted. We never gave a thought to the electricity that flowed silently into our houses, except when a storm knocked it out for a couple of hours. Then we'd complain and feel helpless until it came back on.

I agree, it's better in some ways to make your own music, the way you young people do now. It's just that—most of that other music is lost. Completely lost. Centuries of it. You can't imagine how beautiful it was. I can still summon it in my mind but I can't play it to you.

Almost every day I was in touch with friends and family thousands of miles away: my parents in California, my oldest friend in Sweden, my brother Tom in Nairobi. It was so easy, because of computers and email. We could telephone too, just pick up the phone and hear each other's voices. It cost hardly anything. But there was always the time difference, so email was often better. Now it might not matter so much, since night isn't really night and day isn't really day. But I can't talk to them anymore, time difference or not. I can't talk to anyone who isn't right in front of me, like you. I don't even know if my family is alive.

It seems unlikely, doesn't it? I talk to them in my dreams, sometimes, and wake up sad.

We did manage to see each other, on occasion. We'd save up money and plan a reunion, usually in California. And we'd all fly there. We might grumble about the cost of the ticket, or the annoyances of travel, but we didn't worry about being able to meet again in a few years. Tom once warned us that the time might come when we'd be stuck wherever we were, no more hopping across oceans and continents. I remember that, because although I laughed like everyone else and accused him of hoping he'd get stuck in Kenya so he wouldn't have to come to the next family gathering, his warning bit me like a snake and never quite left my system. Look at us now. He was in Nairobi when the breakdown started and everything unraveled. I wonder what it's like there now. I wonder if Tom's alive and looking back helplessly, like me.

Well, I suppose you're right, it's possible that things are better in other places. We'll never know, will we? They could also be worse. We have each other. We have our little shelters. We have the caves if we have to hide. People lived in them a thousand years ago, did you know that? Long before the white people and the towns.

Where was I?

During the day I worked by myself. My job was rewriting things that other people had written, to make them better. Not a terribly useful job, looking back. I'd studied writing in college. For a while I'd wanted to write novels, but it was too hard.

You like the way I tell stories? Thank you. I'm glad.

Anyway, I was often bored, so I was pleased when Ben interrupted me with a phone call or an email. We'd make a plan for dinner, where to meet or what he might buy on his way home. Sometimes friends would join us. We had a friend called Margaret who was always trying to get us to go to meetings about climate change. Or the war in the Middle East that went on for years, for no good reason. We did go with her a couple of times but it was so depressing. They'd talk about the awful things that were happening, and how the government had caused them or was making them worse. Margaret and Ben and I were almost the only people at those meetings under sixty, it seemed to me. All these gray-haired hippies getting rattled about what was going on but no one knew how to fix it. Sometimes I suspected they enjoyed having something to rail about, it reminded them of their youth. It made me feel hopeless so I stopped going.

No, I don't know what happened to Margaret. I agree, at least she and those others tried. If they're still around perhaps they feel a bit less guilty than the rest of us but their efforts didn't help in the end, did they?

Sometimes, if I had to come into town, I'd meet Ben for lunch at a place where you could sit outside at round wooden tables. There was a sandbox for little kids. They liked children at this place. Each Halloween they'd give gingerbread and hot cider to everyone who came by after the parade. Main Street was closed and all the kids and their parents would walk down the hill in their gorgeous costumes, witches and aliens and insects and skeletons. Hundreds of people together, can you imagine? Ben and I stood on the side of the road calling hello when we saw families we knew. Sometimes we'd dress up too, just to entertain our little friends. One year he was a carrot, a very tall and skinny carrot with twinkling eyes. I was a potato.

Halloween was the very end of October and often it was a mild night, one of the last before winter came. It made me melancholy, the approaching winter. I don't know if you can imagine what it feels like to be very cold. But our house was snug and we had a deliciously warm bed, and after it snowed it was so beautiful, the thick pure white over everything, the blue shadows and brilliant light.

I wish I could take you to that town. Just a main street and a few side streets, small stores, restaurants, a library. A friendly place. You could chat with someone you ran into on the street, or the ladies in the post office. The library building was two hundred years old, a jumble of low-ceilinged rooms with wooden floors. I liked it there. It had a special kind of silence. The books sat on their shelves so quietly, yet you'd pick them up and open them and you'd be in another world, with voices and places and times that you'd never known.

Please don't say that. Don't you think I'd want you to have all this, and more? Don't you think we would have done everything humanly possible to change direction, if we'd known what was going to happen?

Yes. It's true. We did know. We just didn't want to believe it. Maybe you would have acted differently if you'd been around. You'd have been more courageous. If I tried to tell you how sorry I am I'd never find the words. I'd collapse in remorse. What good would that do? Here we are.

Ben usually came home around six in the evening. I'd finish my work and we'd have a glass of wine and cook together. We both cooked. We didn't always do things together. He didn't like dancing. I didn't particularly like going to rock concerts. A mistake. We thought we had years.

Ben.

Please. I can't bear to remember. Do you understand? Of course you don't understand.

Give me a minute. No I do not want a hug.

I'm sorry. I didn't mean to be harsh. It's not your fault. It's my fault, our fault. We let that world destroy itself and we lost everything. Everything. For us and for you as well. I try to look at the future and I can't see that it's ever going to get better.

You're right. The future is not my territory. Maybe you or your children or theirs will change things in ways that I can't imagine and will never see.

Where was I?

The sky.

Once in a while on a moonless summer evening we'd put a blanket out on the lawn and lie there side by side, holding hands, looking up at the night sky. The cats would join us, puzzled but happy to find us out there in the night. The sky was so dark and clear then, we could see countless stars, some brilliant, others barely a smudge. We'd see whatever planets were passing. Sometimes shooting stars. A visit from a comet. Thrilling and beautiful. We can't see the stars any more but they're still there, just as bright. They don't care what's happened to us. They're safe from anything we could ever do.

Ben and I would look up at them and think about how the light of those stars began its journey to our eyes hundreds or thousands or millions of years ago. Each one had its life cycle, billions of years long, but still, a life cycle. Some of those stars were so old, so distant, that for all we knew they no longer existed. Everything ends. And so would we. I felt it as a comfort.

Oh my dear ones. Here we are. Here we are.

## First Day at School

*Or, A New Government Program to Teach Penguins How to Live with Us*

---

Katerina Stoykova-Klemer, *Kentucky*

There is a group of them
waiting for the school bus

Shaking each other's fins
Carrying big books in their backpacks

Playing practical jokes
by stepping on the fat one's coattails

Tying each other's big
floppy shoes

Bringing fresh sushi for lunch
and flowers for the teacher

The teacher prepared the classroom
over the weekend

Sprawled updated maps with pins
to remind us of Antarctica

Put up big welcome signs
and apples on each desk for a snack

## A Small Sedition

Ellen Bihler, *New Jersey*

*Legions of us have hatched and thrived.*
*We fed our queen-mothers well.*
*A thousand virgin queens fly from every nest.*

From behind a hedge,
my neighbor's boy emerges, fat with welts.
I turn away, cored by practicality.
He'll be dead by dusk—
My girl has a few days more.

*There was a time before this frenzy.*
*All labored in the rhythm of day, night, day.*
*There were frays but no war.*

It's the lure of water that brings me out.
The autumn cool's an admonition—
they are foraging too.

*Now we seek the tall walkers*
*even in their beds.*
*Now, our mothers explode with young.*

My beloved used to say,
"Everything in God's world has a purpose,
except the wasps, who are straight from Hell."
I lied like a rose, all through his death.

*Saliva pours from our mandibles.*
*We build our towers. We cover*
*earth and sky with beats of hunger.*

The air itself is a vector.
I've been punctured and marked—
breached by breath,
But the rarest blood defines me.

*Infection flows from our sting.*
*They spread our gift, one to another*
*before they fall.*

I find a birdbath, apple tree, gas grill.
I find a nest, ripe for extermination.
With immunity comes empowerment,
and work to do.

*This poem was written when the author was a sophomore in high school.*

## Tiny Black Rocks

Rachel M. Augustine, *New York*

Black tears run down my face
as I choke and gasp for air
in this black stoned state
because now we fight for breath
and run from the rain
that will burn your skin.
And stare through the bleak black fog,
and wonder
What if my parents had made a difference?
If they had, could I see a flower
that has not shriveled from the black soot?
Could I drink water
that does not taste like dirt?
Could I lay in the sunshine
without getting blisters from the harsh rays,
like before the skies stopped protecting us?
Could I jump in the puddles
and not be afraid of being burned?
It's ironic really.
That the element to control flames can still burn you,
or that earth, that was supposed to be solid and steady,
could create such a problem in the skies,
and the land,
and the life on our planet.
I saw pictures of that life
in the textbooks of our tunneled school.
Of bright birds that sang music
and large cats that stalked the jungle with such grace

and even flying murals.
They were called butterflies I think.
They floated and swooped on the plants
with long tongues and tiny feet.
All I see now are roaches
and rats
under the dim lights in our cave
because the crust is too dangerous
with fire from the skies
and acid from the rain.
But I sneak out at night
and hold my breath
and I look at the stars
and I think
about decisions we made in the past
that mankind condemned themselves
to leave life, and simply exist,
like rats and roaches
under the earth.
Because of tiny black rocks.
Tiny black rocks,
and no will to change.
Tiny black rocks,
and talking instead of acting.
Tiny black rocks,
and denial of the truth.

Tiny black rocks have killed us all.

# Part III
# Revolutions

*Once we have the courage to take in what's going on in the world, the real struggle with global warming begins, in our hearts and minds. How do we respond—emotionally, spiritually, intellectually—when faced with the depth of the damage we have caused? What does climate change mean for our deepest understandings of ourselves and of our place in the world? Faced with these realities, how can we move forward? Do we even want to?*

*Having acknowledged what is happening in the world right now in Part I, and having dared to imagine what might lie ahead in Part II, Part III turns inward, to consider the emotional and intellectual consequences and possibilities of the changes facing us all. Chapter Six, "Twistings," explores explicitly some of the negative emotional responses hinted at in previous chapters—including confusion, guilt, dissociation, fear, hopelessness—unsoftened by the usual calls for action or assertions of hope. Although dark feelings such as these are seldom expressed in public, for fear that they will paralyze people, the premise of this whole book is that this silence is part of what is paralyzing us; only when such negative feelings are expressed can they be confronted, embraced, and worked through, freeing people to begin the process of healing. Then, having moved through the "dark night" of Chapter Six, Chapter Seven, "Turnings," shares and celebrates some of the emotional shifts and spiritual insights through which people are now healing and building new visions and hope for the future. But don't expect any definitive set of moral imperatives or concrete solutions to climate change; those have been widely offered elsewhere. Rather, these authors give us something different: illustrations of people actually taking first, small steps to change their lives and their world, exploring the inner changes required for such steps.*

*In the end, the only answers that mean anything are the ones that you live in your own heart, your own story.*

# CHAPTER SIX

# TWISTINGS

## Late Night News

Malaika King Albrecht, *North Carolina*

We're sleeping. You don't, you won't wake up.
Why did I wake up? Someone calls my name;
someone is calling my name. Am I
awake? Someone is calling me names. We
have too many names and some
we won't answer to. We have much
to answer for.

The world is asleep. How else
can we explain what is happening? The world
sleeps loudly. Like a baby. Someone else's baby
on the page, the TV screen, the ground.

There are many children. The long hallway, darker
than night, is growing longer, is growing more doors
which means more choices, which means I stand
in front of the many doors. Which door,
which child?

*The toddler reaches for his father's*
*casket. The soldier's face is burned*
*into an earless mask, his high school girlfriend,*
*his bride, her own face unreadable.*
*Someone jumps to trade one death for another.*
*This is how you escape a shooter in your school.*
*Mom, would you hide in a locker?*

This place loses mothers, fathers, brothers, sisters,
sons and daughters. I don't know where the bodies
come from or are going, spilling from beds,
from buildings, from every earthly opening.

Our footprints are everywhere, even the moon
and the sea's floor. We have touched the world,
and we are bored. We're in your home now,
reading your email and taking lint samples
from your socks, folding and unfolding your underwear.

The last hiding place of snow will not be here
for long. The polar bears are diving and diving and diving.
Where can we walk in this watery place? Each hurt person
is a stone in our pockets. We are all wounded, trying
to stay awake, treading water. Sing, and
in between songs, hold the dying.

## The Wind

Jim O'Donnell, *New Mexico*

It's the wind that gets me.

It typically comes out of the southwest, although lately it seems like it is coming from every direction at the same time, and it's fierce. The house creaks with every gust and the windows shake in their frames. The west side of the house is forever cold from the wind. Sometimes you can actually feel a draft making its way into the house and it chills you to your spine.

The wind sucks the water from the grasses and junipers and pines and sage. Even the cactus suffers. The wind turns a spark into a conflagration. When it really blows they announce Red Flag days because of the fire danger. We have Red Flag days all the time now. The sky is often so full of dust and smoke and pollen that you have to seriously consider wearing a mask or a bandana. The mountains are masked in haze.

My ex-wife hated the wind. She said it made her brain feel swollen and muddled. She said it confused her heartbeat. She said her electromagnetic pulse stuttered. It was too much for her.

Last August, the wind blew the roof off the house. No kidding. It was a terrible noise. When I climbed up on the roof my stomach dropped; I didn't know that was possible. I crawled back up there on my hands and knees an hour later, dragging a tarp and a bucket of rocks, then huddled the night with my kids in the one corner of the house I could adequately cover with the tarp. The rain poured into our little home. It squirted out of the ceiling lights and fans and I turned off all the electricity. We lay there and watched the lightening outside and listened to our house become a pool.

It's only a fifteen-year-old house, but these houses weren't made for this. The only rainstorm of the year and it floods my house.

I was born and raised around here, and I don't ever remember wind like this. My neighbors, the ones who have been here a very long time, agree. Sure, maybe someone could argue that, technically, the frequency and the power of the wind haven't changed over two thousand years and that all this is within the normal range of variability. Well, maybe. But that makes it all the worse somehow, because I don't remember it like this.

Our community association meets on occasion to discuss our future. We share out some homemade cookies and some organic Guatemalan coffee and Ken from down the road always wonders why we don't have any booze. "That wind will drive a man to drinkin,'" he says.

We live in a sprawling village just south of the Colorado border. There are maybe 3,000 of us scattered over a dozen or so square miles. We're surrounded by public land.

We sit in a circle in our run-down community center and eat those cookies and some organic apples that came all the way from New Zealand. Most of my neighbors want to talk about the road. The paving. The speeders. The culverts. Some people want to talk about "sustainability," about how to come together as a community, and how to support each other in the long haul and in the face of dramatic shifts. Food security tops the list of concerns. How can we all continue to eat in the face of climate change?

"All this talkin' will drive a man to drinkin,'"adds Ken.

Mostly, ideas are met with silence. This isn't a community of deniers, but like most Americans, my neighbors would prefer that this not be happening. When it comes to what should be done, they all seem confused. They'd rather just not think about it.

But the questions are still there, just under the surface. Is it happening? What will it bring? Should you stay? Should you go? Should you hunker down survivalist-style with your guns, or should you take on the much more difficult task of building community? What does it take to create a grocery store for the people? Can we get paid?

It seems that most people around here are resigned to the fact that something big and bad is coming and you've just got to let it come. Then you'll have to adapt. Don't stress about it yet. But I do.

It really is about the food and water. Practical stuff, nothing high-minded. I feel an added burden not only to bring in a salary to support my kids but also to prepare for the day when an income might not buy the food we want or need to eat. It might be too expensive by then, or unavailable. Or not. I don't know. The not-knowing is what I don't like, so I've set down the path of slowly building a compound of sorts where we can become efficient food producers.

And a fence to keep out the wind.

Mind you, this isn't farming land. Our community sits on dry sagebrush flats at 7,200 feet. Summers are harsh, winters are even worse. Still, we have good well water (we think), and the river is less than two miles away. Behind the fence, I planted fruit trees, built garden boxes, and started harvesting water from the roof. Chickens should work too, for eggs and manure and meat. I'm a long way off from being able to feed us three, but little by little, I figure. I'm sure I'm further along than most.

Still, I sometimes wonder if we should move on from here. Quit my job. Sell the house. Find more fertile ground. Should we live closer to, or even with, my parents and my brother's family? Should we build a support network there? What if I can't find a job there?

We've got an emergency store of drinking water in the back closet. There's also a box with white gas, matches, medical supplies, and an ax. We've got enough canned and dried food to last us a month. I never let the gas tank on the car get below a quarter empty.

Am I overreacting? Yes. No. How the heck would I know? I'm as confused as you. We've got one foot in the future and one foot in the past and under that arrangement we aren't going anywhere fast.

I'm trying to talk myself into a gun. I've hunted some in the past. I've got elk meat in the freezer. Mostly I've shot and eaten rabbits and quail. True, there's a river full of fish right over the hill. I'm more fisherman than hunter, I guess.

The young family down the road has decided they are out of here. They've found this island in the Caribbean where food is abundant, the soils are healthy, and the sea is full of fish.

"What about hurricanes?" I ask.

"I've been through a hurricane," says David. "I'd rather face that than this."

"This what?" I ask.

"Exactly."

The forecast for tomorrow: "Mostly sunny, with a high near 55. West wind between 25 and 30 mph. Gusts reaching 40 mph."

Another Red Flag day.

## A Shocking Admission of Heroic Fantasy

Jill Riddell, *Illinois*

My response to the catastrophe of global climate change is similar to the experience I had of 9/11. The whole world remembers what a gorgeous, blue-sky day it was in New York the day of the terrorist attacks, but fewer know that the weather was equally splendid where I was, in Chicago. But unlike the cloudless skies of Manhattan, which filled tragically with smoke, Chicago skies remained serene that day.

Since my husband works downtown, his building and the rest of the tall buildings in the Loop were evacuated. Home by ten o'clock in the morning and wondering what to do, Tim decided to go for a bike ride. He pedaled along the lakefront path for an hour, and later the two of us ate a nice lunch. The two days that followed lacked any semblance of chaos; in fact, they were uncharacteristically calm, since a ban on civilian air traffic meant airplanes weren't on their usual flight paths over our house.

Sure, we Midwesterners were uneasy. Our stores sold out of bottled water and duct tape like everyone else's. The tragedy occupied all conversations, and made work on environmental causes—both Tim's and my professions—seem beside the point. But as it turned out, there was no need for our survival precautions. No terrorist attacked Tim, or me, or our kids. Not that week, nor the next, nor any thereafter. In Chicago, there were no genuine threats.

To readers personally affected by 9/11, I offer apologies for the childishness of what I'm about to say. But here it is: secretly, I envied you. I wanted to have an enemy, a cause, and something to feel unified with other Americans about. I wanted a Pearl Harbor moment. Growing up, I loved books about brave youths who triumphed over polio, escaped from cruel orphanages, or were sightless but plucky. At eleven, I taught myself Braille, hoping perhaps I'd go blind someday and surprise everyone by being ready for it. I craved a worthwhile battle with someone or something. As an able-bodied, middle class, country kid, I hadn't yet found the worthy opponent who could sharpen my skills and give me a chance to prove my courage.

While still in college, I worked as an intern for The Nature Conservancy, and I spent the next twenty-five years trying to make a more biodiverse world. There were many catalysts, both scientific and personal, responsible for my selecting the environment as my issue of choice—one of which was the hope that I might be part of an army that triumphed over something formidable. I hoped I could contribute to something heroic.

Lately, though, I've found myself feeling disappointed. Don't get me wrong, I'm grateful not to have to fight acid rain or the hole in the ozone anymore, but why did they have to fold so quickly? I and everyone else in the environmental movement were so adamant about the threat they presented. Each of these issues, and many others, seemed like worthy opponents, but in retrospect, they improved without my personally doing anything more than rant. I didn't get to sacrifice anything; I didn't get to use Braille.

Then came the crisis of garbage and shrinking landfill space. The 1987 garbage barge enthralled me with its 6,000-mile journey up and down the coast of the Atlantic as it tried to find an open landfill. But America was nowhere near the point of running out of open space into which we could shove our garbage. It turned out the unmoored barge was the result of clumsy bureaucracy rather than scarcity of land.

When global climate change came along, I permitted myself cautious optimism. Could this be the enemy that would require the cunning of a war zone journalist and the dignity of a hurricane survivor? Would I need to make painful sacrifices? Could I tap deep reservoirs of strength and stalwartness?

I saved my pennies to buy a Prius and energy-efficient appliances. This included a state-of-the-art European furnace so futuristic that no one in the Chicago region knew how to fix it when it shut itself down, which it tended to do whenever the outside temperature dropped below freezing. It *was* darned efficient, though, when it worked. (I say all this in past tense because we had to scrap it and get a more ordinary boiler a couple years later. The efficient one went to the metal recycler.) Tim and I invested heavily in the insulation of our house, the dullest purchase ever made. We had an energy audit done. We purchased carbon offsets for our family, and gave them to our relatives as Christmas presents. I bought an additional offset for our family vacation, even though it was already covered by the first.

If ever there was a "global warmist," it was I.

To date, though, I haven't seen or felt the results of my efforts, nor do I have any direct experience of climate change. I'm not in edgy Manhattan, which the famous map in *An Inconvenient Truth* showed to be in peril of flooding. I'm not in Florida or Calcutta, which will sink into the sea by the end of the century, nor am I in the western U.S., which has already seen a two-degree rise in temperature.

While climate models predict that it's a virtual certainty that Chicago's temperature will eventually rise two degrees, this puts the city at the climate of Champaign, Illinois, which is a shift subtle enough that most folks won't notice. At the wicked end of things—if we continue our hedonistic ways along with everyone else around the globe—Chicago could end up with temperatures like northern Texas or southern Arkansas. This would be lousy for our natural ecosystems, but not so catastrophic for humans; millions of people already live happily in such a climate. The inconvenience Chicagoans are likely to suffer isn't torrential windstorms or withered aquifers, but rather an onslaught of people. With our moderate climate and abundant supplies of fresh water, Chicago's future isn't filled with tarps and duct tape—just higher real estate values.

I'd like to be in a position to write an essay about climate change filled with rage and despair. But that sort of drama belongs to the more deserving. I'm learning to be content with making unheroic, everyday changes and encouraging the rest of my fellow earthlings to do the same. These actions won't lead to glory. But if enough of us do it anyway, they might lead to success.

## The Watcher

Susan Palmer, *Colorado*

I was born when television was just beginning and there was no such thing as anti-lock brakes. All my sixty years, I have watched the human civilization in the United States grow and change and change again. I have seen the ways of tribes who live off the land in far off places and those who depend on the sea for life. I have imagined myself a part of such cultures and sighed with delight that my life is so easy, that food can be purchased, that warmth is controlled by the touch of a finger and a bit of power, that my water is clear and drinkable.

Around 1990, at the beach in the summer, I noticed the sun hitting my skin like it was shooting straight pins into me. I stopped enjoying the beach. The hole in the ozone layer was in the news, but that didn't frighten me … it seemed just an interesting event.

Around 2001, I began to have dizzy spells whenever the sun had a major flare. Now *that* was more interesting, since it affected my personal well-being. Still, I was mostly an observer, allowing the dizziness to be for as long as it wished to be (usually only an hour).

Around 2005, I had accumulated enough input to realize that the climate all over our world was dramatically changing. Bears coming out of hibernation too early, gulls lacking food fish in the far north, glaciers melting, deserts forming, pines catching diseases, tornados in unusual places, unexpected heat waves. I was impressed, excited.

I will be here to see the end of our easy way of life. Crops will fail, forests will change, and wildlife will either migrate or die back. Insect populations will alter, and birds will move to friendlier areas. Deserts and plains will be born and transformed. And when crops fail and water patterns shift, humans in North America will probably panic and become aggressive toward neighbors—or maybe helpful, we'll see. Beauty will be ignored in a single-minded obsession with sustaining a lifestyle that cannot be sustained. But there *will* be beauty for those of us who are the watchers, who don't plan on struggling to stay alive at any cost.

When water no longer comes out the tap due to power failures or lack of sources, what will you do? No power means no heat. What wood will be available to people in Kansas? Who knows today how to make fire with no matches? The trucks will bring no food. Stores will be emptied. How will you cook?

I will watch. People will panic and become vicious. What use would it be for me to store water or food? Others, stronger people, will steal it from me, will hurt me or kill me to get it. Fine. I will store water enough for a short time. I will share it with anyone. When it is gone, I will drift away into death more peacefully, not interested in the grand struggle to survive ... survive for what? For war and illness? For pain and fear? No, better to leave with a smile on my face, admiring clouds and the way dust flies down the lane.

I am over sixty years old and do not have small children to fret about. I have left my body three times, due to fevers of one sort or another, so I know for certain that I am more than my body. I have been lucky to have had a vivid experience of one of my own past lives, so this too reassures me that I am not my body, stuck with this life, this world.

What is the point in living in anguish and fear for the last part of my life when I can instead concentrate on joy and beauty? I will be the watcher as long as I can.

## Hopeless For Today

Helen Sanchez, *Montana*

Climate change: What am I thinking? How am I feeling? Every day is different from the one before it—which is kind of like the climate; ha, ha. In all seriousness, I'm seriously depressed about it today. Mostly how I respond is to ruminate. Sometimes I ride my bike to work instead of drive. One reason I do so is because I don't want to get fat. Another is that I don't have to buy fuel. Last but not least, it feels good to be a "part of the solution," as they say. I feel great, in fact, until I get stuck behind a car or a truck with a bad muffler.

Although human beings are not 100% responsible for climate change, we surely have helped it along. I feel about it the way I feel about murder, rape, genocide, ethnic cleansing, child abuse: I feel hopeless and full of despair. I feel like no matter what I do, it will never be enough to make up for the damage my existence causes, let alone the damage caused by our entire species since the Industrial Revolution. Even my vegan diet contributes to global warming with all the methane gas it produces from my body.

The space I occupy could be used by a family of moles, say, or a couple of geese. Let me sacrifice my life for theirs, instead of the other way around. I contemplate the quickest method of suicide—an Earth-friendly method, of course. A bullet through my head would be efficient, but I'd have to buy a gun from Wal-Mart. Think of all the sweatshop labor I'd be supporting. I can see the little children in China right now, assembling parts for my gun.

Maybe I could open a vein. I already own a kitchen knife. No shopping, no hassle. But let's face it; I don't have the guts for that sort of thing.

This is what I do about climate change: I feel bad about it. I think bad thoughts over it. But I don't really *do* much about it, other than the cycling and the vegan diet. I don't protest. I don't lobby, or write my congressman. I don't even pray about it. I just kind of hope it resolves itself. If that means our human race blows up in a puff of nuclear smoke or something, then so be it. I hope it happens as quickly and painlessly as possible.

Long live the cockroaches, who are sure to survive it all.

*The following was written as an exercise for an introductory English course in college.*

## Wrath of Human upon Gaia

*In Response to Assignment 6, "Creative Writing with Analysis"*

Quynh Nguyen, *Florida*

**I. Poem**

I want to apologize for my actions.
My lack of consciousness; for creating
the flames around me.
To apologize for my passion
to feel the heat that warms my needs
but burns my children.
I want to repent for my desire
to eat the green and swallow the token seeds.
I want to redeem myself for feeling pleasure
from the charring pain.
I want to apologize for my actions before I leave.
My needs have led me to a virgin.

◊ ◊ ◊

**II. Analysis**

This poem discusses the destruction we have created upon our world.

*I want to apologize for my actions.*
*My lack of consciousness; for creating*
*the flames around me.*

"I" represents the human race and our society. I use this narrative to portray my own guilt and my understanding that I am a part of this society.

The flame around us is the hell that we have created for ourselves. This can be taken in a religious aspect, with the amount of sins that we have committed when we pollute or devastate this Garden of Eden. This hell is also global warming.

*To apologize for my passion*
*to feel the heat that warms my needs*
*but burns my children.*

This "passion" refers to the enjoyment that our destruction provides us. This negative effect on nature gives us satisfaction that makes us forget the wrong that we are doing. It is our addiction to material goods and living habits that are not ecologically beneficial. To keep with the theme of hell, it burns our children, the generation of the future. We are willing to burn our children to satisfy ourselves. Humans have a selfish drive that many may attribute to survival skills.

*I want to repent for my desire*
*to eat the green and swallow the token seeds.*

"Green" represents the crops and animals that the earth provides for us. It is the natural course of things for birds to spread the seed, for bees to pollinate. It is the way that things have been set up for us. But we have consciously decided to swallow the seed, stopping the natural order of things, and prevent any hope for the future. We swallow this because it satisfies our hunger. Green can also be seen as the food that feeds our money culture.

*I want to redeem myself for feeling pleasure*
*from the charring pain.*

This goes back to our pleasure that comes with every destructive move we make as humans. The charring pain is on ourselves, yet we enjoy it because we are ignorant of its effect. We are overwhelmed with the gratification we receive from our own actions. The charring and burning is a way of describing the killing off of our human population, as we destroy ourselves amidst this hell that we have created.

*I want to apologize for my actions before I leave.*
*My needs have led me to a virgin.*

We have a pattern of leaving when things get rough. Our apologies represent the green consumers that will try to save the world as long as it accommodates their lifestyles. Many Catholics, for example, believe that they can repent for their sins and be given admission into heaven; this concept is also evident in the way we try to "save the environment." Our needs have led us to move from this barren land and find a new place where we can implant our seed and cultivate a new manmade world. We have come to accept the arrival of an apocalypse but have convinced ourselves that modern technology and human intelligence will allow us to live on after the dying of Gaia.

# CHAPTER SEVEN

# Turnings

## Strand Sonnets

Kathryn Kirkpatrick, *North Carolina*

*1*

We took highway 52 to the beach,
a supple two-lane road flanked by fields of corn
and tobacco, crops inching up and up each
mile we left behind our rolling hills. We scorned
the interstate that day, those featureless
hours. I wanted to see the actual land.
And as we headed through the long slow reach
of lowland, I felt my grief eased and borne

by the green expanse. My sadness turned small
on that empty, forgiving road. It was here,
still, the local and particular, not yet
paved and owned by men who'd never see all
that love and care could do in just one place, near
enough to know, land befriended and well kept.

*2*

Land befriended and well-kept was all we met
that day. Silos of grain, fields of strawberries.
I wanted to see the land because I felt
bereft. In taking another country
my country had been taken. I couldn't yet
bear miles of fast food neon, that ugly,
garish sprawl built like a bad movie set.
I'd traveled four-lane 74, harried

by traffic lured by the bait of the franchise.
That long assault of signs and snarl of stores
now seemed backdrop to a recent craven war.
And so I sought another landscape, wiser,

more various, unplundered, through main streets more
to scale with human needs for love and care.

*3*

The human need for love and care won't be met
by one more McDonald's off the Interstate.
For years I'd head into the standard routes, set-
jawed, filled with dread. Then books on tape
appeared to compensate for boredom and wretched
food. It was all bearable, the raped
land, our stolen hearts, if narrative could let
us go elsewhere, beyond our present state

of numbed servitude to cars, to ease, to greed,
and not always our own. As if the rich
were gods we might sate so that they'd do nothing
worse than this: score the land with shopping strips, feed
us what we cannot taste and does not nourish,
pave the present until it gives us nothing.

*4*

And if the present gives us nothing
we'll always wish to be somewhere else,
not in the bowels of Wal-Mart, pricing
products no one made with love, lined on shelves
by those who cannot make a living.
They are the ones who spend themselves
so that we're saved half the price of everything.
And yet to be there in that broad, crass store sells

us too. As I did not feel bought and sold
driving back roads to the beach though I bought
butter beans from the road stand where a farmer
ate soul food, the collards soaked in beer, and
I ate a peach where I stood when I ought
to have waited. I could not wait for later.

*5*

But wait. Later's only now in retrospect,
so if we aren't here now we won't be then.
That farmer's face is with me still, respect
for the broad hands that worked his few acres, when
all around the BiLos and Winn Dixies neglect
his harvest, buy the crops of other men
a thousand miles away. Shall we say he lacked
ambition because of what he failed to own?

Harvesting machines. More land than he could ever
know except to say the value of in cash.
It must be how a parking lot gets made—
that no one mourns the earth sealed up and severed.
What happens there beneath the tarry ash?
What happens there when everything is dead?

*6*

When everything is dead, we walk across it.
Arrive and park. Depart. That used up space
is useful as long as we're not in it
much, the hoods of cars such shiny carapaces,
I dream them on their backs like beetles. It
helped to think them helpless when I traced
a route from home to work, sixteen without
a car. I'd walk beside the six-lane, braced

against the heat and speed, then turn into
the parking lot, an asphalt plain that wavered
as I walked. Half a mile of all the same
except the day a car slowed and I saw too
much, a leering man with his pants unzipped,
exposed, like me, in that waste without a name.

*7*

In that waste without a name we travel
aimless. Tar, gravel, and sand. Then overpass,
exit ramp, barren land. Still we travel,
our tongues stuck to the roofs of our mouths, guileless
and dumb. We travel, the questions we will
not ask lodged deep. *Who among us has*
*the heart*—tar, gravel, and sand—*to ask? Will*
*you be among those who grow so slowly wise?*

When Eisenhower went to Germany,
he fell in love with Hitler's autobahn.
Tar, gravel and sand. Overpass, exit ramp,
barren land. The interstate, a fascist's dream,
forced speeds, forced routes, forced food, and market cons.
*Whose trash is this? Whose greed? Whose fast food dump?*

*8*

I never saw a trashy fast food chain
that day we drove the back roads to the beach.
And I arrived without my heart in chains,
released, without the need to ask each
wave and stretch of sand for more that might sustain
me. I was already full, full enough to teach
my wounds the grace of healing, enough to train
each hour on the trellis of each day, each

moment to give back as much as I had taken.
The tide advanced and I attended. The gulls
and pelicans arrived. The waves frothed and
exclaimed. And when the land gave way to ocean,
I saw we could own none of it at all,
not earth or sea or sky or bay or sand.

*9*
Neither earth nor sea nor sky nor bay nor sand
belong to anyone in parcels or in tracts
except to steward and to love. Mt. Pleasant
women told me that they never tell the facts
about where sweet grass grows, but rather tend
and share so that there always will be baskets,
Gullah baskets, amaretto sweet. And
when I stop at stalls on the beach road so that

I can compare the prices, find a deal,
I find there is no deal, no one willing
to take less, no one eager to have more.
A daughter, hair coiled tight, won't let me steal
her mother's work half-price. She turns the spiraling
weave, a low flame in her eyes, shows me the door.

*10*
And I am taught by the low flame in her eyes.
The second time I stop I understand
the value of this work. I'm slowly wise,
enough to cradle baskets in my hands,
to pay the proper price for every hour of life
invested there. I learn these woven strands
need misting to keep them supple sweet, surprised
the long grass smells of ocean meeting land.

How would it be to live with fewer things,
but each one made like this, from stories carried
deep through many lives? A pattern handed down
and changed. A color and a shape that sing
through grief. Our lives restored to beauty, we'd
awaken, our forsaken hearts finally found.

*11*
When our forsaken hearts are finally found
we'll peel back parking lots like bandages
from giant wounds and find beneath a sigh, a moan.
Released from all that tarry ash, unbound,
we'll need some time for grief, to sit down
with the startled land, and like the man who manages
to pray after long silence, to mourn what has been done
of which we were a part. Our cries amid the ashes.

And then what will we plant? How will we begin?
Before my father died he had been drawing
a family tree, the names in his block hand.
On one side of the page, his mother, embedded in
her Irish clan, and on the other, another string
of Irish names, except one read: part-Indian.

*12*
*Part-Indian* it reads. I hold it out,
the family tree my father drew, look to see
beneath the watermark, in his block hand, how
she was born in Indian territory,
my father's father's mother. Without
a name, a mystery on the family tree,
she calls across the miles and years about
what I have lost, what we have lost, the stories

now beneath our feet. And when my father died
we buried him in the usual way,
a casket lowered into set concrete,
no pine box as he'd asked, no swift return. I
wonder if he lies there still, away
from land he loved. I wonder if he's free.

*13*

My father loved the land he knew with wonder.
He liked to say that god lived all around us,
in roses, swallows, hickories and thunder.
And in us too. If he were here, he'd trust
this surf to bring up pompano, cast farther
out for no more than we'd need, enough, just
enough. If there was greed in him I never
saw it. There was only sadness, wide as

any gulf, a grief so deep, how could it be
that it was his alone? Perhaps we can't pave
over the memories of a conquered people.
So great-grandmother, send up a shoot in me.
I take my share of grieving, but save, yes save
the strand I'm meant to weave. And make me equal.

*14*

Make me equal to the work of grieving
so that I can return for what I left,
those strands of sweetgrass that I need for mending.
I want the heart to name the wrong, the theft
of what we love. I want the heart for mending.
Sweetgrass smells of almonds and the heady heft
of ocean meeting land. Back to the rending,
that's where we'll find the heart made wholly deft.

How can we know the public from the private
wound? How can we heal the one without the other?
I think perhaps my father's grave may keep
now that the concrete splits and earth is sated,
now that his bones become the branches of the birch,
his song the mourning dove's before she sleeps.

*15*

When we took highway 52 to the beach,
land befriended and well-kept was all we met.
Our human need for love and care was met, each
acre gave each hour back, a present.
In retrospect, I think I tried to reach
back to the dead, across the years to what
seemed wasted and somehow find a name, teach
myself to look beyond the fast food chain at

earth and sea and sky and bay and sand.
And I was taught by the low flame in their eyes,
the women whose hearts are woven with sweetgrass
into baskets, African. Part-Indian,
my father taught me love and grief, and I,
through grieving, am made equal to the asking.

## Beyond Denial

Willow Fagan, *Michigan*

A few winters ago, Michigan had an unseasonably warm December, and a lot of people I knew—my friends, my mom, the local bloggers I read—joked, "Maybe global warming's not so bad after all." These jokes were, of course, a way of managing our deep fear; we were encountering the unsettling prospect that climate change was *here*, no longer a grim specter beyond the horizon but among us, all around us in the air that blew too warm.

But in the summer, when even the cool evenings are warm, and the green is everywhere and exuberant, I forget the lessons of winter. I walk around the tree-lined streets, or look out my window at the bright grass, and it seems impossible that the natural order of the world is threatened. The carefree easiness of summer seems to promise that fall will never come, let alone the unknowable new season of global warming. The broad-leafed trees, the grass, the birds and squirrels and spiders are the same as they have always been, the same as they were when I was a child, as far back as I can remember. This naïve faith is the same species of emotion as my childhood certainty that nothing truly awful would ever happen to me or to the people I loved.

As a queer person, I have some experience with denial. I would like to be able to tell you that emerging from denial is like moving from blindness to light, prison to flight, as a butterfly escapes her cocoon. But the truth is more messy, more cyclical. I catch glimpses of the destruction that our collective actions have caused or might cause, of what global warming means or might mean, and then I retreat back into distraction, into disconnection, into false promises of safety. Do butterflies ever return to the broken shapes of their former shells, in need of shelter from snow or storm? Do they long to?

The shapes of human-wrought shells are longer lasting. The

ones we build in our minds can endure long past their usefulness, stubbornly refusing to change shape even when the world they are meant to reflect has radically transformed. In my own life, my parents handed down to me their inherited certainties about sex and love, sin and God, and I had to dig through their fear and my own to discover the shape of my own sexuality, the color of my own true desires, right for me; same-sex, queer desires, as it turned out. This revelation was the first shattering of the safe bubble of my childhood dreamworld. For me, the choice came down to this: understand a core part of my being as sick and wrong, or abandon the worldview I had held my entire life, the worldview shared by my parents, my teachers, and most of my friends, in which God reigned over the Universe like a benevolent king, and all our stories had happy endings in Heaven. I chose to leave that faith behind.

This may seem, perhaps, to have little to do with global warming. And yet it is the history I always speak from, the experiences which color my perceptions, whatever they might rest upon. I have found myself drawing on the skills and courage and wisdom I drew on then, as I rebuilt my understanding of the world in relative isolation, as I now face unpredictable changes that once again threaten the order of the world: the order in my mind, the order out there, the spinning circle of the seasons.

That spinning circle has become an important part of the order within my mind. I rediscovered the sacred in the natural world, in the shapes that tree branches make against the sky, in the swooping shadows of bats at dusk, in the constant motion of rivers and streams. I now walk an earth-based spiritual path, and one of the oldest, most central symbols of my path has been the wheel of the seasons, an image I use to center and ground myself, to open myself to the Divine, to the unfolding process of life, death, and rebirth—as exemplified by seeds sprouting and expanding in spring, fruit ripening in summer, leaves falling in autumn, roots slumbering beneath bare, frozen branches in winter, to wake in green splendor once again in spring—around and around, a spiral, the slow periodic quickening of the Divine.

Now, though, that eternal cycle appears to be a wheel wobbling, threatening to spin off its axis. How can I face this? How can I bear the thought of all that I hold sacred being irrevocably harmed or even extinguished? The web of life unraveling into dust and silence?

There was a gap in between when I rejected Christianity and when I found a new spiritual home in earth-based religion, in particular the movement that some of us call neo-paganism. This time of uncertainty was a terrible openness, like being naked and alone on a plain which harsh winds had swept clean of anything but dust and grey. I did not know what to believe, or *how* to believe, how to know that anything was true, reliable, solid.

I feel threatened by the return of such scouring doubt when I try to imagine the future of global warming. Just as my discovery of my queerness revealed many of the comforting structures of my childhood to be sources of damage and deception, so too the full accounting of climate change indicts so much of what seems commonplace—cars, suburbs, oranges and bananas in northern climes in the middles of winter—as dangerous objects with an aggregate destructive power worse than that of a nuclear bomb.

Yet facing the truth of my identity has led me to a deeper appreciation for my life, a new way of tasting the skin of the world, a new way of being at home in my own skin. I trust that a similar transformation can occur through confronting global warming.

This is how I cope now, how I bear the threat of eco-apocalypse:

I bear it in small doses.

I bear it in the company of others whose hearts are rooted in earth, whose eyes tear up at the cutting down of trees, the mangled bodies of deer and raccoons hit by cars, the news of the ending of an entire species.

I bear it by grieving, by allowing the swell of pain and tears. When this flow is dammed, I fear tsunamis but I open my heart to my heart, again and again, *and I have yet to drown.*

I bear it by reminding myself of what remains, the ongoing beauty of the world.

And at the worst of times, when I fear that nothing of the beauty I know will remain, I stretch to encompass this truth: The wheel of the seasons has never been eternal. There was a time before the Earth existed at all; and there was an Earth before the seasons, when all was live magma and there was no sea and no clouds. If we view the cycle of the seasons as a symbol larger than its actuality—as all real symbols are—then what it symbolizes, that deep order of endless change, of death and renewal intertwined and inseparable, may still be real even as the seasons themselves go haywire. The cycle of change is the heartbeat of the Universe—expansion and contraction, in and out, in and out, the rhythm within each body, each star, each atom, pulsing.

The massive destruction that global warming threatens to unleash sometimes seems boundless, endless. Yet, if the worst comes to pass in a tragedy far beyond words, there is an order so much vaster, so much older and deeper and wider than we can comprehend, which nothing we do can threaten.

In the worst of times, I take comfort in that.

## Glooscap Makes America Known to the Europeans

Sydney Landon Plum, *Massachusetts*

One summer, my children and I were camping in Acadia National Park, Maine, and it was raining. Making the most of the park's offerings, we tucked into an alcove of the very small space that is the Abbe Museum in Bar Harbor, watching an animated film created by grade school students of the Passamaquoddy nation. It was there that I was introduced to Glooscap, the hero-creator figure of the Abenaki people, as he and Grandmother Woodchuck set off in a stone canoe to sail across the sea. In the words of Charles Leland's published version, "This was before the white people had ever heard of America. The white men did not discover this country first at all. Glooscap discovered England, and told them about it." Closet iconoclast that I am, I loved the perspectival shift this story proposed.

Glooscap—also spelled Gluskabe, Glooskap, Gluskabi, Kluscap, Kloskomba, or Gluskab, and pronounced *glūs-kă-bā*—is always great: in size, in manner, and in his beneficence. He is so big that his head touches the stars. He created, among other landforms, the Minas Basin and the Five Islands in the Bay of Fundy in Nova Scotia—and if you know the Bay of Fundy, you know that this is a big deal. He is a teacher, as well as a transformer and creator. He has a sense of humor, but he is not a trickster. He has a well-developed sense of justice, but sometimes, in his duty to protect the people, he acts in haste and has to revise his giant works. I like a heroic figure who can be wrong, who shows some humility in the face of some force greater than he.

As with all Native stories, the tales of Glooscap come down to us through oral tradition, each storyteller (including those grade-school filmmakers in Maine) adding their own style, insight, and humor to the communal well of culture. One such story, Glooscap's creation of the Basket-tree People, is retold by Abenaki storyteller and author Joseph Bruchac in the collection *Rooted Like the Ash Trees: New England Indians and the Land.*

Glooscap was lonely, "so he decided to make the people." Since he worked mainly in stone, he tried that first, but the people he made of stone were heavy, awkward and slow; they damaged the plants as they stepped on them. Then he saw the beauty of the ash trees, so he decided to make the people from them. When he shot arrows into the trees, humans stepped forth, who "could dance like the ash trees in the wind, they were graceful and their hearts were green and growing."

The people who lived along the coast and into the interior of what we know as Maine called themselves Wabanaki, or Abenaki, "people of the dawnland." Two of the member tribes of the Abenaki are the Penobscot and the Passamaquoddy, who share their names with the rivers marking their territory. To the north of the Passamaquoddy River lies the territory of the Mi'kmaq and the Maliseet, who were probably separate nations before the wars of the eighteenth century but became part of the Abenaki confederation after that time, and in any case are part of the larger Algonkian language group that includes the Abenaki as well. All of these peoples have stories, and Glooscap figures in quite a few.

In one Mi'kmaq story, "How Glooscap Created Sugarloaf Mountain" (retold by Elder Margaret Labillios, among others), the people called upon Glooscap to do something about the very big beaver that had built a dam across the Restigouche River in New Brunswick. The dam prevented the passage of the salmon upriver to spawn. The people were aware that if the salmon could not get up the river to have their fry, there would be no more salmon, and the people would have no food in the winter. When Glooscap hit the dam it broke into two parts, which became Heron Island and Bantry Point. Glooscap killed one of the giant beavers, swinging it around by its tail then letting it go so that it landed miles away, where the beaver carcass turned to stone and became Sugarloaf Mountain. Then Glooscap stroked the heads of the other beavers, and with each stroke they became smaller and smaller until they reached the size they are today. Glooscap promised the Mi'kmaq that the beavers in New Brunswick would never again grow big enough to build a dam that could cut off the rivers so the salmon could not get through.

There is a parallel legend about a giant beaver told by the Pomuctuck peoples of Massachusetts, another Algonkian tribe. In this version of the story (as recorded by Deacon Phinehas Field in the late 1800s), the Pomuctuck call upon the spirit hero Hobomock because the beaver is eating people. Hobomock kills the beaver, and the animal's body is used to create another Mt. Sugarloaf, this time in western Massachusetts (near Deerfield). This legend is part of the folklore of Lake Hitchcock, the massive lake that once existed in the Connecticut River Valley.

In *Wisdom of the Elders: Sacred Native Stories of Nature*, scholars David Suzuki and Peter Knudtson admonish their readers not to view the stories of indigenous peoples, or the people themselves, as part of a romanticized past. Rather, traditional Native ways of knowing are often just as empirically grounded and ecologically insightful as are Western scientific perspectives.

*Native knowledge about nature is firmly rooted in reality, in keen personal observation, interaction, and thought, sharpened by the daily rigors of uncertain survival. Its validity rests largely upon the authority of hard-won personal experience—upon concrete encounters with game animals and arduous treks across the actual physical contours of local landscapes, enriched by night dreams, contemplations, and waking visions. The junction between knowledge and experience is tight, continuous, and dynamic, giving rise to "truths" that are likely to be correspondingly intelligent, fluid, and vibrantly "alive."*

Indeed, the Glooscap stories speak about places that are significant in the natural history of New England and Maritime Canada. Along with being cultural treasures to those of us intrigued by the awe-inspiring tides of the Bay of Fundy, within most of the stories there is an experiential core and a connection to some aspect of place that science has described in the intervening years. Moreover, within each story is an emotional truth about our relationship with the nature that shapes our lives today in the twenty-first century.

The giant beaver, *Castoroides ohioensis*, was a huge species of rodent—up to eight feet in length and of an estimated weight of nearly 500 pounds. It is not known if these beavers built dams, for their teeth were different from the beavers we know, and their tails were rounded, not flat. They lived in North America, and fossils have been found near Toronto and around the Midwestern United States (in Ohio, Minnesota, Indiana, and Wisconsin). The megaflora they favored were an aspect of the Eastern woodlands, also. Evidence points to the giant beaver becoming extinct approximately 10,000 years ago, concurrently with the mammoths and mastodons.

In both the Mi'kmaq and the Pomuctuck versions of the story, the confrontation between the hero and the beaver represents a conflict between two cultures—one human, the other animal—shaping their environments. The human way of living by the river and the beaver's way cannot both be sustained; something has to change. The historical core of these stories is confirmed by the findings of modern geology and archaeology, which tell us that the giant beaver were disappearing from the landscape at the same time that human life was coming into it. Both changes were the result of climate change, over which neither people nor *Castoroides ohioensis* had any control. Landforms changed; beavers became smaller and did not compete with people. The people who inherited a land with smaller beaver gave credit to Glooscap (or Hobomock) for creating an environment in which they could thrive.

In another Passamaquoddy story, "How Glooscap bound Wuchowsen, the great Wind-Bird, and made all the Waters in all the World Stagnant" (retold in Leland's *Algonquin Legends*), Glooscap is with the men who go out in their canoes to fish. But it is very windy: "[I]t grew worse; at last it blew a tempest, and he could not go out at all. Then he said, 'Wuchowsen, the Great

Bird, has done this!'" Glooscap searches out the Great Bird and confronts him for having no compassion for his grandchildren, asking him to go a bit easier. The Great Bird replies, "I have been here since ancient times; in the earliest days, ere aught else spoke, I first moved my wings; mine was the first voice,—and I will ever move my wings as I will." Glooscap, confronted by the immovable forces of a disinterested universe, ties up the Great Bird and stuffs him down between the deep rocks. The winds stop. In fact, the dead calm created in the absence of the winds makes the waters stagnant. Glooscap cannot paddle his canoe; the people cannot fish. Glooscap is made to realize by the people that this is a catastrophe, of his making. He goes back to release Wuchowsen—but only unties one wing, so that "the winds have never been so terrible as in the old time."

This, too, is an example of a Glooscap story paralleling an environmental event for which there is scientific evidence. After the end of the last Ice Age, there was a "cold snap" referred to as the Younger Dryas stadial. The great ice sheets were melting and enormous quantities of cold water were pouring into the North Atlantic. Some theorize that this influx of cold water may have caused the Gulf Stream to stagnate, causing a period of cold in Europe and the North Atlantic portion of North America. Perhaps what the Glooscap story describes is indeed the stagnation of the Gulf Stream in the North Atlantic—as a result of the disappearance of Wuchowsen.

For a while at the beginning of the twenty-first century, it seemed that the story of Wuchowsen might be told again. The increasing flow of fresh water into the far North Atlantic from melting ice caps has the potential to slow the Gulf Stream once again, which according to some scientists would mean that New England and Maritime Canada would grow much colder and might, eventually, be covered by ice. There would be substantial loss of nonhuman life in this area I call my home. There would probably be substantial loss of human life; certainly there would be a significant migration. Everything I most value—from Cadillac

Mountain in eastern Maine to the Museum of Contemporary Art in western Massachusetts—would be obliterated.

When this theory of a coming ice age first circulated, I was darkly infatuated. It was the story I am often tempted to tell: the only way to save the earth is to get rid of (most of) humanity. We are the creatures who foul our own nests, who can't be bothered to learn how to live within our means so that our children, and our children's children, will have the means to live. I grew up during the decade following the destruction of Hiroshima by nuclear weapons, when the possible eradication of humankind was a common storyline and when my schoolmates and I practiced hiding under our desks or waiting in a hallway for the world to end. For many years the destruction of everything I knew and loved seemed like a given. Why shouldn't it be ice, instead of fire?

And yet, how can I love the way the nuthatch hops headfirst down the tree trunk; the way the pasque flower offers a second, ghostly bloom; the sound of the great tide of the Bay of Fundy—and not want all this to endure, with humankind nearby to participate in crazy life? I have finally come to see that the desire for the annihilation of the human race is a product of both cynicism and laziness—as it would mean avoiding the hard work of trying to find a better, sustainable fit for human life in the various ecological niches we inhabit. Slacker mentality. In any case, the more recent studies have shown that even the cooling effects of a slower Gulf Stream cannot offset the rising temperatures caused by other changes to our climate. It seems we will not perish in ice, but in fire after all.

The Glooscap stories suggest that the Abenaki people in the past—like ourselves—experienced environmental changes that affected their livelihood and were sometimes at odds with non-human life forms over resources. All these years later and we still have the same issues. Another important aspect of the Glooscap stories is the centrality of meddling with natural forces. We meddle with the places where we live in order to survive: to have food and shelter and to create future generations. We also meddle with the environment because we can. In coming to terms with the changes to our planet, we must take into account the continued presence of humanity and their continued messing about with systems that probably worked better once.

The fishermen who called upon Glooscap in the face of present disastrous changes in the environment looked for a quick fix. We do not have Glooscap, but we have technology, and constantly rely on technological meddling to quickly remedy complicated problems. Glooscap's first efforts were as catastrophic to the environment and the people as the dangerous situation he was sent to remedy. I am particularly wary of technological quick fixes, as these seem to flow from the same chemistry sets and slide rules that created a mushroom cloud to darken my childhood and threaten our living planet. We have to do better.

Finding the right action is difficult, in part because we have not been able to sort out the complexity of the issues confronting us. We have more than one story of what is happening. Many of the stories told about climate change are factually and spiritually misleading—yet people listen to them anyway. There is no one person telling a strong story with a clear voice. There is no consensus about what to do. There are many people watching the tides and waiting for a story, with different needs and agendas, which makes consensus all the more difficult. It would be nice to have a hero like Glooscap. It would be wonderful to have the technology to bind one wing of Wuchowsen (but only one wing!) and put the waters into their right relationship with the land and the fish and the people. Perhaps that will happen yet, but it is not something I can help with.

We are at a moment when very different stories may be heard across divisions of time, space, and culture. In *Into the Forbidden Zone*, novelist William Vollmann interviews survivors of the tsunami that devastated Fukushima, Japan, and damaged nuclear reactors. He tries to describe the multiple dimensions of this natural and manmade disaster, in order to ask how many more times this might or must happen. One person he interviewed said that he did not question putting reactors at Fukushima because it was the government's decision. On the other side of the world, on the Bay of Fundy Explorer website, one writer wonders why the Fukushima disaster has not led to increased questioning of the exposed site of the Point Lepreau Nuclear Generating Station. Could something be gained for both communities if these people could tell their stories to one another? In *The Boston Globe*'s photo essay on the wildfires caused by the drought in Texas, the last photograph is one taken from the International Space Station, showing smoke plumes across east-central Texas—streaks of grey-white against the buff curvature of land stretching from the Gulf of Mexico and beyond. What if people begin to understand that events in their own backyard are big enough to be seen from space? On The Poetry Foundation website, poet Jennifer Fitzgerald reads a poem she wrote about the aftermath of the storm called "Sandy": sunlight on ruined houses, the hope that the landscape will somehow return to what it was before the storm, a barefoot woman refusing shoes because someone else might need them more. What if all the people who are without shoes were able to tell their stories and be heard by those whose actions might lead to more sustainable practices?

I am placing a great deal of faith in storytelling, but, like meddling with our planet, we humans have been doing it for a very long time. It is hard work to create effective narratives combining diverse voices. It is difficult to create an audience and to get people to listen. We need compelling narratives—stories of attachment to nature, like the Glooscap stories—not stories of destruction and world's end.

All these centuries later, Glooscap yet again may set off in his canoe to tell stories of the land to new generations of "discoverers."

## Credo

Charlie Krause, *Maine*

Grabbing my splitting maul as the sun rose over the pond against the backdrop of the fall trees that had lost their leaves I thought to myself that the best way for me to help the earth was to become part of it by living a good life that was both environmentally friendly and sustainable in that while I would have to use the same basic raw materials as most people to get by in this crazy world I would make the effort to use less, recycle more, waste little, and whenever possible live as close to the earth as I could by growing my own food, building my own structures to live in with local hand-cut wood, and applying the new technologies of wind power and solar power to my lifestyle to make as little impact as possible on the world after I am gone and to improve the world at large while I am alive and able to make a difference in the direction I believe the human race needs to go as it moves forward into this new century where things like oil and nuclear power have become the enemy and pollution is rampant across the globe because of the human race's need for big cars and houses and plastic replacements for almost anything that used to be made by the hands of people who had no television or radio to look at and listen to and thus learned to make things with their hands and to keep themselves warm with their labor and ingenuity and in the end we all should know that it will always be better to split wood than atoms.

## Doing Work, Causing Change

Monica Woelfel, *California*

It's an unseasonably warm October day in Soquel. As I carry a tote bag of damp laundry out to the line, I try not to think of global warming. Instead, I hope that the day's heat is simply an example of seasonal variations, ones that have always come and gone.

My grandmother was born not far from here. She grew up in a dusty gold town. She hung laundry to dry in this very same way, with a similar weight in her arms of the damp fabric, so solid it feels like a living thing.

I set the tote bag on the ground, carefully arranging it so it won't tip over and dump my clean white sheets into the dust. That happened last week, and the dust didn't come off easily, not without another washing.

I lift a pillowcase from the tangled wad of clothes and pin it to a metal clothesline with a wood clothespin. Each piece will need to be lifted carefully, so as not to toss others in the dirt, and hung between two clothespins. It's time-consuming to handle each item in this way. I could have—I still could, I think—throw the whole mass into the dryer and push a button. There. Done.

I resist.

I lift out socks—one, two, three, four. I pair them and secure each pair to the line like the salted mackerel I saw once in the Shetland Islands, rows upon rows of fish hung on lines to cure.

I pull a pair of jeans from the pile next, dangle them upside down from their hems, and pin them firmly. Behind me, a pair of hummingbirds buzz and chirp. They dodge about through the flowers like crazy wind-up toys. I have to take a deep breath and let it out slowly to quell my impatience. I notice the sun warming my shoulders. It feels good. I slow down.

A breeze comes out of the redwoods and bay trees below this clearing. It smells cool and spicy of damp places, creeks and hollows in the valley. The breeze lifts the clothes I've hung so far. It fills them the way wind fills a sail.

In college, I studied for a semester on board an oceanography ship, the *RV Westward*. The *Westward* wasn't one of those huge Jacques Cousteau–style research yachts, but rather was (and still is) a 100-foot schooner, a graceful, slender sailing ship. As students, we learned oceanography—peering for hours into a drop of seawater at radiolarian that looked like elaborate blown-glass ornaments—and we learned to sail the ship.

We started from St. Thomas in the Virgin Islands, traveled southwest to the island of Bonaire, just off Venezuela's coast, then headed northwest to Belize, and finally through the Straits of Florida to Miami. The circuit took us six weeks, with time off to re-provision and to explore during our stops.

Toward the end of our trip, I was assigned night bow watch. As we came in toward the orange smudge of light that marked Miami on the dark horizon, I scanned the waters around us nervously; the first mate had impressed upon me how much oil tanker traffic streamed through these straits. As a small craft, compared to the massive tankers, we had to be alert; while technically the rules required *them* to avoid *us*, since they were under power and we were not, the reality was that we showed up on their radar screen—if we showed up at all—as a tiny, easily-missed dot. Especially easy to miss if an engineer happened to be tired or had had one too many beers with dinner.

Looking out for oil tankers, I got to thinking about Americans' voracious appetite for oil. These tankers run day and night all around the world—as do pipelines and trucks and trains—to feed our energy needs. Like nearly every other American, I owned a car. I had always considered oil a necessary evil. How else would we get around?

I had learned in physics class that some energy must be expended to gain motion. I recalled the equation $E=1/2\ mv^2$, in which the mass of a vehicle, m, requires energy, E, to get it into motion. V equals velocity. I thought of motion, therefore, as an exchange: the burning of natural resources—such as coal or uranium or petroleum—in exchange for travel. The resulting pollution poisons us—along with plants, animals and the planet itself—but at that time, while I mourned the cost, I considered it unavoidable. Either we pollute, I thought, or we go—as oil company execs are fond of saying—"back to the Dark Ages," when most people didn't stray far from home. Those who did, rarely made it back.

Yet, straddling the schooner's bowsprit that night, I felt as if a curtain slid aside on the Petroleum Wizard of Oz. It struck me that this ship, with its thirty students, three mates, three scientists, and a captain, had just made a 2,600-mile journey and hardly expended any resources whatsoever. How could that be? What had moved us all this way?

The answer, of course—as any five-year-old could have shouted out—was the wind. The wind had pushed us along in the direction it was going anyway. Nothing had been burned, nothing wasted. No effluent was created, no heat, no smoke, no radioactive waste. The only recognizable work had been when three or four crew members had gathered together to share the job of hauling the mainsail into position to catch the wind—singing rousing sea chanties as we did so, another non-petroleum energy source.

Conservative scientists and mainstream thinkers have a tendency to call environmentalists' solutions Pollyannish. Yet there's nothing foolishly optimistic about traveling great distances without creating noise or pollution, without depleting resources. A sailboat is simple, beautiful to the eye, and—to apply a phrase scientists like for describing a system or equation that works in a surprisingly effective way—it's as "elegant" as they come.

A kind of peaceful delight filled me that night. It *is* possible to live our lives and not destroy our home, I realized. We just need to rethink the ways we go about things. It didn't have to be a question of going "back" to some uncomfortable, rudimentary way of life but rather of going forward, using both old and new knowledge to craft a future that, in its ability to allow ourselves and other creatures to live healthy lives, will make *today* look like the Dark Ages.

When I think of the Dark Ages, here's what I see: the bubonic plague sweeps through Europe. A woman kneels on a dirt floor by an open fire pit. Smoke blackens the walls; a shank of meat drips its fat hissing into the fire. I picture sooty hands, flea-ridden mattresses, and sheep and chickens and cattle resting just outside the door, if not inside. People everywhere are dying gory, agonizing deaths. The world as they know it is ending, and none know if anyone will survive.

Today, global warming numbers show a future in which none of us may survive, with the possible exception of cockroaches and bacteria. The world as we know it appears to be ending.

The solution to the bubonic plague, it turned out, was staggeringly simple: cleanliness. More specifically, the edict, "Don't live with rats." It's amazing to think that all of that suffering could have had such an easy remedy. Could it be that the cloud of destruction hanging over us today has an equally simple solution?

As I watched the wind off the Strait of Florida fill the sails of the *Westward* that night, I knew that it could.

My pillowcases billow, then deflate on the clothesline. The two hummingbirds get so caught up in their mad flight that they barely miss my head. They land in a plum tree next to the line and chitter at me as if the near-collision were my fault. I laugh.

I wonder if my grandmother found the task of hanging laundry pleasant, if it was for her a chance to slow down and enjoy the sun on her shoulders. I'm surprised that a task I resisted with such impatience turns out to be the antidote to the very anxiety that troubles me. The "medicine" of taking time to hang my laundry in order to save energy has ended up offering the unexpected cure of relaxation and pleasure.

I hang my last piece of wash, a blue hand towel. The whole process has taken ten minutes. I think about the electricity I would have used in the dryer, one of the biggest energy hogs of household appliances. I consider how this laundry, hanging here peacefully for a few hours, will get just as dry.

I spend a few minutes in the open air, slowing down enough to pin each piece of wash to the line. The effort or energy expended will be the sun's and the wind's and mine. Otherwise, no resources will be consumed, no pollution created.

The breeze rises again from the valley. I think of the mainsail on the *Westward*, billowing full. I think of elegant solutions.

## How to Be a Climate Hero

Audrey Schulman, *Massachusetts*

One afternoon a few summers ago, I was on a commuter train when I heard someone yelling behind me. I didn't pay attention because I was breaking up a fight between my kids. I figured the noise was from some college students having fun. The third time the person yelled, I turned around.

It was a boy, about six years old. He was standing on his seat, screaming, "My mom's having a seizure." The only part of his mom I could see were her legs sticking out into the aisle, convulsing. Arrayed around the train car, staring, were forty other people, mouths open. Not one of them doing a thing.

Humans tend to freeze like this—the bystander effect, as it's called. The phenomenon was first demonstrated in 1968 in a famous psychology experiment by John Darley and Bibb Latane. For the experiment, the subject was asked to fill out some forms. He or she assumed that these forms were necessary information preparatory to the experiment, but in fact the experiment had already begun. While the person circled multiple-choice answers, smoke began to sneak out of a vent in the room—thick smoke, grey smoke, the kind that says *fire.* The experimenter timed how long it took for the subject to leave the room to find out what was going on.

The single variable was whether there were other people in the room. If the subject were alone, 75% of the time she or he would leave the room inside of a minute. But if there were others in the room working away at their papers—actually actors paid by the experimenter to stay there, heads down, pencils working, ignoring the smoke—the subject stayed there with them, 90% of the time. Stayed there filling out forms until the smoke was too thick to see through—until if there had been a fire, it would have been licking at the walls.

In the decades since that first bystander experiment, it's

been repeated with many variations on the type of emergency: staged robberies, lost wallets, people in the hallway crying for help, etc. Every time, if there were more than one person witnessing the event, all of them were almost certain to do nothing. In some of the later bystander-effect experiments, the subjects have blood pressure cuffs on and what they say is recorded. Their pulse races, their blood pressure rises. They mutter "shit" and "holy hell." From their reactions, it's clear that they recognize what's happening as an emergency and feel great urgency about it. Still, they stand there, frozen.

Remember this fact: although we feel safer in a crowd, that's actually where humans are most incapacitated. The bigger the crowd, the stronger the effect.

On the train, the boy was loudly identifying this as a true emergency, his mother physically demonstrating the urgency of the matter—and still everyone sat there, mouths open. Half of them had cell phones clipped on their belts, but not one of them dialed 911. No one ran to get the conductor. While they watched the woman convulse, each of them glanced around and believed everyone else must be sitting still for a good reason. Perhaps the others had some inside knowledge, that this was a movie being filmed or a scam being tried or that the kid was playing some sort of mean joke.

Each person thought that if this were real, then surely with forty other people here there must be someone who knew how to deal with seizures. There must be someone competent, with professional training and a medical vocabulary. Each person assumed, "I should be the last person to help. I don't know dinky about seizures."

Right now, everyone understands that something horrible is happening to the planet's climate. The heat waves and forest fires, the floods and droughts. But there's seven billion of us now—quite the bystander effect. So we stay in our seats filling out forms, working dutifully, trying to ignore the smoke swirling thicker around us. We mutter under our breath, our hearts race, while we wonder why no one else is doing anything. Like the adults on the train, passively watching as a child screams for help for his mother, each of us bustles about our normal lives, feeling increasingly uneasy about the shifting climate, but assuming that it couldn't be as bad as it seems because then surely everyone would be marching in the street about it. We all seem to be saying to ourselves, "If climate change *were* real, then there must be someone better than *me* at getting people to demonstrate against it. I don't know dinky about activism."

On the train with the epileptic mother, I got to my feet for two reasons.

One, I knew about the bystander effect, I had studied it in school and written about it before. Knowledge about how badly humans react in emergencies is the best way to short-circuit the effect. Research has shown that as long as you remember this tendency of humans to passively gawk, you are inoculated against it. In fact, simply by reading this essay, you are much more likely in the next emergency you encounter to get out of your seat and do something.

The second reason I didn't sit still is that I'd experienced the bystander effect in the past. As a teenager, I'd found a lost puppy sleeping in a park. It was maybe three months old, with a pure white coat and pink tongue. A friend and I patted it for a few minutes before it bolted away from us, out into the street. There was a car coming. Let me be clear: the car was not that close, and I easily could have stepped out, holding up my hand to stop it, or simply scooped the puppy up and walked away to safety. Instead, both my friend and I stared, as passive as if watching TV.

As the car got closer, in that elastic moment of fear, I learned something about myself—that I could be a small scared person, that I could passively watch harm happen to something defenseless. I didn't like that feeling. Although the car skidded to a halt in time and the puppy was OK, I never wanted to see myself behave that way again.

On the train, hearing the boy yell, I didn't wonder why everyone stayed still. I knew why. I stepped forward, yelling out, "Someone call 911. Someone get the conductor. Anyone here have medical training?"

Then, something fascinating happened. Before I moved, everyone's faces were contorted with terror, as though they were the ones having the seizure or as though this woman thrashing around like a dying fish might start biting their ankles. But as soon as one person did something, telling them what to do, how to help, the fear in their faces melted away. Two other people stood up to help. Four others whipped out their cell phones to call 911. One person ran for the conductor. They just needed someone to break the group cohesion and start the action. They desperately wanted to do good. Like me with the puppy, while they'd stared at the woman convulsing, their assessment of themselves had been rapidly plummeting. They didn't know why they were frozen, but they were beginning to grasp the possibility that they might live out the rest of their lives knowing they hadn't done a thing while this kid screamed louder and louder.

I cradled the convulsing woman's head so at least she wouldn't thunk it hard against anything. Two other people tried to reassure the boy. The conductor stopped the train and we waited for the EMTs. By the way, it turns out that aside from getting people to start moving, I wasn't terribly useful. I remembered reading that during seizures it's important to make sure the person doesn't swallow their own tongue and suffocate on it, but after this incident, I found out that the swallowed-tongue thing is a myth and that the most you should do is to make sure they don't hurt themselves slamming their body parts around. However, in the moment, without this information, I determinedly tried to jam my fingers between her grinding teeth to grab hold of her tongue. The point is not to be the most competent person—which I am definitely not—the point is to get people moving. Anyone can do that.

Psychologists know a lot about fear: how it starts, how it changes over time. If a person experiences fear for long enough, especially if there's no perceived way to fight the danger, the fear shifts into anxiety and depression. In a famous experiment by Martin Seligman, dogs were caged up and then repeatedly electrocuted through the metal floor. The shock was hard enough to hurt, not kill. The shock was preceded each time by a bell being rung. After the bell there was nothing these dogs could do but wait for the pain. After a few days of this experiment, the dogs lay down and whimpered—not just when the bell rang, but all the time. They wouldn't eat; they wouldn't take interest in other dogs. They basically acted like they needed a lot of Prozac and a straitjacket. That whimpering puddle of depression is called "learned helplessness."

Seligman had a second group of dogs that had a safe room inside their cage. This room wasn't electrified. When those dogs heard the bell, if they jumped super-quick for the safe room, they could possibly avoid the pain. These dogs never lay down and whimpered. They ate normally and functioned. Yes, they sometimes didn't reach the other room fast enough and then they got shocked and it hurt like hell, but the pain wasn't the point. The point was that they had a sense of power in the world, of agency. They felt active and capable of defending themselves. They weren't sitting frozen in their seats with no idea of what to do.

A few years ago, when my first child was born, I became paralyzed with fear about climate disruption. It was so clear that our children would be punished for what we adults were doing to the world. My child would suffer for our sins and there was nothing I could do about it. I got depressed. I got anxious. I lay on the floor whimpering for a while—metaphorically. Then, from sheer desperation, I started writing letters to editors. Finally, one of my letters—in support of Cape Wind, the proposed large wind-turbine farm off Nantucket on the Massachusetts coast—was published in the *Boston Globe*. Soon after, the head of Cape Wind, Jim Gordon, called me up personally to thank me. The thrill I got. The sense of agency.

After that I was out of my seat. I believed that there was a safe room I could at least try to get to, if I moved super-quick. Now I go to every demonstration. I write to every politician. I insulate my house fanatically. I don't own a car. Every year I do a little more: composting kitchen waste, buying at farmers markets, recycling, buying secondhand. Using carbon calculators, I've figured out that I lowered my family's emissions 50% in seven years. That's a big step. Because of my actions, my fear for my children's future is not incapacitating. I'm not depressed. I'm striding down the aisle trying to help. I'm learning as I go. Not only have I improved my emotional state, I've broken group cohesion and started to pull others from their seats. I've gotten friends and relatives to insulate more and drive less, to admit the problem and to start thinking about solutions.

By the time the EMTs arrived, the woman had stopped convulsing and was breathing easier. I was still holding her head, and I've got to tell you, a human head gets heavy after a few minutes. They woke her up to lead her, dazed, from the car. The boy trailed after her. Every one of us passengers called out to him he'd done a great job. We told him he'd saved his mom. That group that had been so scared and frozen a few minutes ago was now grinning and relieved. We were slaphappy with love for other humans and ourselves.

Each of us knew that the situation could have turned out so differently.

The scientists tell us that Americans must lower our carbon emissions at least 80% by 2050 to avoid the worst effects of climate disruption. Let's imagine the year is 2050 and we've managed to lower our emissions enough. As I've already seen in my own home, radically decreasing emissions is not so hard. Surely the U.S., the most innovative and wealthy nation in the world, can do a lot more in forty years than I did in seven. Let's imagine we've gotten out of our seats. We've strode down the aisle. We've done our best with whatever information we had. Whether or not we incompetently tried to grab at slippery tongues, we still broke the bystander effect. We got the country moving. We didn't lie down whimpering and depressed. Filled with our own sense of agency and our communal effort, we grin around at each other, proud of humanity and ourselves, slaphappy with love for our planet.

We are already at 400 parts per million of carbon dioxide in the atmosphere, the point at which the scientists say the Earth has a 50/50 chance of shifting to a new climatic system. Because the planet is so large and unwieldy, the climate takes a little while to shift. We can balance here for a few years before the dice are cast. Every indication, from ice caps to defrosting tundra, seems to show that this, right now, is the tipping point.

This is our moment.

The kid on the train is standing up and screaming for help. The weather is convulsing. We are all staring. Perhaps you never thought you'd get a chance to play hero. Here it is. Let me tell you, you'll feel better. As soon as you get out of your seat, much of the fear and depression will go away. Others will follow. It's so much easier than you can imagine.

It doesn't matter if you aren't sure what to do. Make your best guess. Call the EMTs. For God's sake, get the conductor.

# Coda

## The Lucky Ones

Penny Harter, *New Jersey*

Chicxulub crater, Yucatan peninsula, sixty-five million years ago:
The asteroid plunges into the sea.

A huge mushroom cloud,
blossom of steam and fire,
of dust and gas, annihilates all life
a thousand miles around.

Deadly tidal waves rush inland.
And now a deadly fall of Earth-born meteors,
launched on impact, rains down to ignite
the world's forests, their holocaust spewing
ash and smoke to shroud the sun,
freezing the Earth.

Flora and fauna,
ammonites to dinosaurs—
half Earth's species dead.

Yet we are here,
our mammalian ancestors
the lucky ones.

We are here, the lucky ones,
our impact spawning floods and toxic smoke,
spreading wave after wave of annihilation
as if we were grooming the planet,
each time pulling out more fur
from the human comb.

## The Angels Are Rebelling

Barbara Crooker, *Pennsylvania*

The angels are rebelling, descending
from on high, with flashing swords
and terrible wings. Their gaze
is pitiless, as they fall from the broken
sky. There is no sanctuary, not even
in cathedrals among the effigies
and tombs. Gaia has sounded the alarm,
hammering, hammering on her golden bells.
The fields are on fire. The pale horse and rider
traverse the land. In iron and ink,
the final story is about to be written,
as the last of the glaciers slips into
the sea. The bells are ringing.

## About the Authors

**Malaika King Albrecht** ("Late Night News") is the author of three books of poetry, *What the Trapeze Artist Trusts* (honorable mention in the Oscar Arnold Young Award), *Lessons in Forgetting* (a finalist in the 2011 Next Generation Indie Book Awards), and *Spill*. She is the founding editor of *Redheaded Stepchild*, an online magazine that accepts only poems that have been rejected elsewhere. Malaika lives with her family in Ayden, N.C., and is a therapeutic riding instructor.

**Rachel M. Augustine** ("Tiny Black Rocks") currently attends the State University of New York College of Environmental Science and Forestry. While in high school in Clifton Park, New York, she was president of both of the school's environmental clubs and served as a student ambassador to Russia. Rachel's writing has won her scholarships, recognition, and several funded international and national trips (including to China), and she hopes to author a book in the future.

**Kristin Berger** ("Learning Their Names as They Go") is the author of a poetry chapbook, *For the Willing,* and was coeditor of *VoiceCatcher 6*. Her essays and poems have appeared in *Calyx, The Blue Hour, Mothering, New Letters, Passages North,* and *The Pedestal Magazine*, among other publications, and have been nominated for the Pushcart Prize. In 2012, Oregon State University's Spring Creek Project awarded her an Andrews Forest Writers Residency. Kristin lives in Portland, Oregon, and curates her blog, *Slipstream,* at www.kristinberger.wordpress.org.

**Ellen Bihler** ("A Small Sedition") is the author of two poetry chapbooks, *Late Summer Confessions* and *An Avalanche of Blue Sky*. Her poetry has appeared in *Cream City Review*, *American Journal of Nursing, Journal of New Jersey Poets*, *International Poetry*

*Review*, and elsewhere, and she won two honorable mentions in *New Millennium Writings.* Ellen is a Registered Nurse working with severely disabled children, and a Certified Advanced Clinical Hypnotist.

**Jamie Sweitzer Brandstadter** ("The Innocence of Ice") was born and raised in Glen Rock, Pennsylvania. A graduate of Mansfield University and the Bread Loaf School of English, she teaches English at Dover Area High School. She lives in Dover, Pennsylvania, with her husband Josh and their dog, John Denver.

**Dane Cervine** ("The Last Days") is the author of *How Therapists Dance* and *The Jeweled Net of Indra.* His poems have been recognized by Adrienne Rich, Tony Hoagland, *The Atlanta Review,* and *Caesura,* have appeared in a wide variety of journals including *The Sun, The Hudson Review, Catamaran,* and *Red Wheelbarrow,* and have been featured in various anthologies, newspapers, video, and animation. Dane is a therapist in Santa Cruz, California, and serves as Chief of Children's Mental Health for Santa Cruz County. His website is at www.DaneCervine.typepad.com.

**Barbara Crooker** ("The Angels Are Rebelling") is a poet whose works have appeared in a wide variety of magazines and anthologies as well as in ten chapbooks and four full-length books. She is the recipient of numerous awards and fellowships, among them the 2007 Pen and Brush Poetry Prize and residencies at the Moulin a Nef, Auvillar, France, and at the Tyrone Guthrie Centre, Annaghmakerrig, Ireland. Barbara grew up in the mid–Hudson Valley of New York and currently lives and writes in Fogelsville, Pennsylvania.

**Alan Davis** ("What We Say When We Say Goodbye") was born in New Orleans but now teaches English and creative writing at

Minnesota State University, Moorhead. His third collection of stories, *So Bravely Vegetative*, won the Prize Americana for Fiction 2010, and his work has appeared in *The New York Times Book Review, The Hudson Review, The Sun*, and many other print and online journals.

**Todd Davis** ("On the Eve of the Invasion of Iraq") is the author of four books of poetry, most recently *In the Kingdom of the Ditch,* along with one chapbook and numerous scholarly publications. He is a winner of the Gwendolyn Brooks Poetry Prize and teaches creative writing, environmental studies, and American literature at Penn State University's Altoona College.

**Julie Dunlap** ("Annapolis Bus Ride") is coeditor of the anthology *Companions in Wonder: Children and Adults Exploring Nature Together* and an award-winning author or coauthor of numerous children's books, including *John Muir and Stickeen: An Icy Adventure with a No-Good Dog*. She earned a Ph.D. in Forestry and Environmental Studies from Yale University and coordinates a schoolyard habitat grant program for the Audubon Society of Central Maryland.

**Margarita Engle** ("Search") is the Cuban-American author of *The Surrender Tree,* which received the first Newbery Honor ever awarded to a Latino/a. Her other young adult novels in verse include *The Poet Slave of Cuba, Hurricane Dancers, The Firefly Letters, Tropical Secrets,* and *The Wild Book.* She has received two Pura Belpré Awards, two Pura Belpré Honors, three Américas Awards, and the Jane Addams Peace Award, among others. Her most recent novel in verse is *The Lightning Dreamer*. Margarita lives in central California, but also can be found at www.margaritaengle.com.

**Willow Fagan** ("Beyond Denial") is a writer of speculative fiction, with occasional forays into poetry and creative nonfiction;

his work has appeared in *The Year's Best Science Fiction and Fantasy 2011* and *Men Speak Out: Views on Gender, Sex and Power, 2nd Edition.* He is also a healer-in-training and a Reclaiming witch. Willow grew up in southeastern Michigan and currently lives in Portland, Oregon.

**Diane Gage** ("Ursus Maritimus Horribilis") is a writer and visual artist living in San Diego. Her poems have appeared in publications such as *Puerto Del Sol, Seattle Review,* and *Phoebe* as well as in two chapbooks, *THAT Poem, Etc.* and *Mother Dreaming.*

**Lilace Mellin Guignard** ("The Darkness") earned an M.F.A. in California and taught tenth-grade English in Appalachia before studying literature and environment at the University of Nevada at Reno. Her poems and essays have appeared in journals such as *Ecotone*, *Hawk & Handsaw, Sundog, ISLE*, and *Orion Afield.* Lilace currently lives and writes in rural Pennsylvania.

**Margaret Hammitt-McDonald** ("Trees of Fire and Rust") is a naturopathic physician in private practice and an adjunct instructor at the National College of Natural Medicine in Portland, Oregon. Prior to her medical career, she earned a Ph.D. in English from the City University of New York and taught literature and composition courses at Bronx Community College, and she still teaches rhetoric and composition at Clatsop Community College. Margaret writes poetry, science fiction, and the occasional essay, and lives with her spouse of 18 years, their daughter, and eight rescued cats in Seaside, Oregon.

**Penny Harter** ("Blue Sky" and "The Lucky Ones") writes in Mays Landing, New Jersey, and teaches in the NJSCA Writers-in-the Schools program. Her most recent books are *One Bowl* (a prizewinning e-chapbook), *Recycling Starlight*, and *The Night Marsh*, and her work has appeared in numerous journals and anthologies worldwide. Penny's awards include the William O. Douglas Nature Writing Award, three poetry fellowships from the New Jersey State Council on the Arts, and a 2010 residency at Virginia Center for the Creative Arts.

**Marybeth Holleman** ("Thin Line Between") is the author of *The Heart of the Sound: An Alaskan Paradise Found and Nearly Lost* and *Among Wolves: Gordon Haber's Insights into Alaska's Most Misunderstood Animal,* as well as coeditor of *Crosscurrents North: Alaskans on the Environment.* A Pushcart Prize nominee, her essays, poetry, and articles have appeared in dozens of publications and on NPR, and she has taught creative writing and women's studies at the University of Alaska, Anchorage. Marybeth also runs the Art and Nature blog at www.artandnatureand.blogspot.com, and more on her work may be found at www.marybethholleman.com.

**Kathryn Kirkpatrick** ("Strand Sonnets") is the author of six collections of poetry, including *Unaccountable Weather* and *Our Held Animal Breath*, and the editor of *Border Crossings,* a collection of essays on Irish women writers. She earned a Ph.D. in interdisciplinary studies from Emory University and currently holds a dual appointment in the English department and in the Sustainable Development Program at Appalachian State University. Kathryn lives with her husband, William Atkinson, in the Blue Ridge Mountains of North Carolina.

**Charlie Krause** ("Credo") lived off the grid for seventeen years, in a hand-built house powered by solar panels. He currently works as manager and head chef of the student center at Unity College. He is also a self-proclaimed "outsider" artist in a variety of mediums, and with his wife Barbara Walch runs an organic farm producing food and flowers in Thorndike, Maine.

**Tara L. Masih** ("Be Prepared to Evacuate") is author of *Where the Dog Star Never Glows: Stories* and editor of *The Rose Metal Press Field Guide to Writing Flash Fiction* (a ForeWord Book of the Year) and of *The Chalk Circle: Intercultural Prizewinning Essays* (a Skipping Stones Honor Book). She has received a finalist fiction grant from the Massachusetts Cultural Council and *The Ledge* Magazine's fiction prize. She lives in Andover, Massachusetts, in a new home that doesn't flood. Her website is at www.taramasih.com.

**Kathryn Miles** ("To Wit, to Woo") is the author of *Adventures with Ari, All Standing,* and a forthcoming book about Hurricane Sandy. Her articles and essays have appeared in publications including *Best American Essays, Ecotone, History Magazine, Outside,* and *Terrain.* She is a member of the MFA faculty at Chatham University and a scholar-in-residence for the Maine Humanities Council.

**Benjamin Morris** ("The Last Snow in Abilene") is the author of numerous works of poetry, fiction, and nonfiction, with recent work in publications such as the *Oxford American*, the *Southern Quarterly*, and the *Tulane Review*. A member of the Mississippi Artist Roster, he is the recipient of a poetry fellowship from the Mississippi Arts Commission and a residency from A Studio in the Woods in New Orleans. More information is available at benjaminalanmorris.com.

**Golda Mowe** ("A Jungle for My Backyard") was born and raised in Sarawak, on the island of Borneo, to an Iban mother and Melanau father. After graduating from university in Japan and enduring ten years of corporate life in Sarawak, she ended her career to explore the cultural traditions and modern experiences of Borneo's indigenous people, guided by her own memories of childhood evenings spent in the longhouse. Golda is the author of the novel *Iban Dream* and of numerous stories and articles, available at www.gmowe.ws.

**Quynh Nguyen** ("Wrath of Human upon Gaia") majored in International Business and Economics at Rollins College in Winter Park, Florida, where she also published articles in the *Rollins Undergraduate Research Journal.* Originally from Vietnam, she arrived in America at the age of 4 with her mother and older sister; in 2007 she traveled back to Vietnam, which provided her with a global perspective on the current social climate. Since graduating in 2011, Quynh has worked in various business positions for a municipality, a consultation firm, and other financial companies both in the private and in the public sectors.

**Jim O'Donnell** ("The Wind") is the author of *Notes for the Aurora Society* and *Rise and Go.* A freelance writer/photographer and former archaeologist, he tells stories that show the link between human beings and the ecosystems around them. He writes, walks, fishes, and marvels from northern New Mexico. His travel writing and photography can be found at www.aroundtheworldineightyyears.com.

**Susan W. Palmer** ("The Watcher") is a grandmother who creates art, poetry, and short stories from her home near the Colorado Rockies. She has worked for years with poets and writers in her community, and her own work has appeared in several print magazines, three chapbooks, and the e-novel *No Tribe of His Own.*

**Sydney Landon Plum** ("Glooscap Makes America Known to the Europeans") is the author of a book of natural history essays, *Solitary Goose*, and editor of *Coming Through the Swamp: The Nature Writings of Gene Stratton Porter.* Her works of poetry and nonfiction have been published in *Prairie Schooner*, *South Dakota Review*, *Organization and Environment*, *ISLE,* and elsewhere. An adjunct instructor in English and creative writing at the University of Connecticut, Storrs, Sydney divides her time between western Massachusetts and mid-coast Maine.

**Jill Riddell** ("A Shocking Admission of Heroic Fantasy") lives and writes in Chicago, where she teaches writing at the School of the Art Institute. Her work has appeared in *Chicago Wilderness, Garden Design,* and other magazines, and in the online project *Acupuncture after the Apocalypse.* Jill won an Audubon Award for excellence in environmental reporting.

**Roxana Robinson** ("Snowshoe Hare") is a novelist, essayist, and biographer who writes frequently about the natural world. She is the author of nine books, most recently the novel *Sparta,* and four of her books were named Notable Books of the Year by *The New York Times.* Her work has appeared in *The New Yorker, The Atlantic, Harper's, The New York Times,* and elsewhere. She has received fellowships from the NEA, the MacDowell Colony, and the Guggenheim Foundation. Roxana divides her time between New York and Maine.

**Jo Salas** ("After") is the author of *Improvising Real Life: Personal Story in Playback Theatre,* an account of an innovative theatre approach in which personal stories are told by audience members and enacted during a performance. Other published work includes short stories, personal essays, articles, and chapters about Playback Theatre, and a bilingual collection of stories told by immigrants. Born in New Zealand, Jo currently lives and writes in the Hudson Valley of New York.

**Helen Sanchez** ("Hopeless for Today") graduated from the University of Montana in Missoula with a Bachelor of Fine Arts. Born in Washington, D.C., she also has lived in Seattle, New York City, San Diego, and La Paz, Bolivia, and has recently moved back to Montana.

**Audrey Schulman** ("Edged off Existence" and "How to Be a Climate Hero") is the author of *The Cage, Swimming with Jonah, A House Named Brazil,* and *Three Weeks in December.* Her writing has appeared in *Ms., Grist, Orion,* and other periodicals. She lives near Boston with her family and runs an energy-efficiency nonprofit called HEET.

**Paul Sohar** ("Winter Visions") is a poet, translator, and writer with ten books to his full or partial credit, the most recent of which is *Silver Pirouettes.* His work has appeared in numerous publications such as *Kenyon Review, Out of Line,* and *Writers Journal.* He lives in Warren, New Jersey.

**J.R. Solonche** ("Polar Bears") has been publishing in magazines and journals since the early 1970s. A four-time Pushcart Prize as well as Best of the Net nominee, he is coauthor of *Peach Girl: Poems for a Chinese Daughter* and author of *Beautiful Day.*

**Harry Smith** ("About the Weather") was the author of 16 books of poetry and essays. His literary press, Smith Publishers, produced more than 70 titles over five decades, and he was known for his support of the literary arts (especially small presses) as a founding member of COSMEP and as creator of the Generalist Association. He received the Small Press Center's Lifetime Achievement Award and PEN's Medwick Award for his epic poem, *Trinity.* A long-time resident of Brooklyn Heights, New York, Harry lived most recently in Cape Elizabeth, Maine, with his wife, playwright Clare Melley Smith, dogs Teddy and Monty, and cats Dusty Roads and Jasper. He died in Portland, Maine, in November 2012.

**Katerina Stoykova-Klemer** ("First Day at School") is the author of three poetry books, most recently *The Porcupine of Mind.* Her poems have appeared in publications throughout the U.S. and Europe, including *The Louisville Review, Margie, Adirondack Review,* and others. She is the founder of poetry and prose groups in Lexington, Kentucky, and hosts *Accents,* a radio show for literature, art, and culture on WRFL, 88.1 FM, Lexington. In January 2010, Katerina launched Accents Publishing.

**Carla A. Wise** ("Burning to Zero") is an environmental writer on a range of topics, from agriculture and forest management to endangered species and climate change. She has a Ph.D. in biology and worked as a conservation biologist before giving up science for writing. Her essays and articles have appeared in publications including *The Oregonian, High Country News, The Huffington Post*, and *The Utne Reader*. Carla lives with her family in Corvallis, Oregon.

**Monica Woelfel** ("Doing Work, Causing Change") studied oceanography as an undergraduate through Sea Semester and at Swarthmore College before earning her M.F.A. in creative writing from the University of British Columbia in 2003. Her fiction, nonfiction, and poetry have appeared in a variety of publications, including *The North American Review, Sierra Magazine,* and *The Sun.* Formerly a Washington State Artist in Residence, Monica now lives in Sacramento, California.

## Acknowledgments

I want to express my deepest gratitude to everyone who responded to the original call for submissions to this project. Your insight, care, and courage confirmed my sense that there was, and is, a complex and continuing conversation about the emotional and spiritual dimensions of global warming that so far has taken place largely beneath the public debate. Even if your submission is not included in this volume, your ideas and writing were essential in giving shape and support to the project overall; in a very real sense, you helped to make this book happen.

At various stages of the project—from initial conception to final product—I have been engaged, challenged, and buoyed by my conversations with a number of friends and colleagues. My thanks to David Adams, Cynthia Baker, Lundy Bancroft, Robert Buehler, Larry Buell, Melissa Burrage, Sue Chandler, Julie Crockford, John Elder, Joey Fox, Elaine Hatow, Soren Hauge, Kurt Hoelting, Ann Holmes, Brian Holmes, Keith Holmes, Curt Meine, Karla Merrifield, Seth Mirsky, Carolyn Norton, Chris Pastorella, Joe Pavlos, John Pavlos, Margaret Pavlos, Katie Payne, and Elizabeth Skidmore.

I also would like to remember here two people whose lives ended during the course of this book's creation: contributor Harry Smith and my father, Fred R. Holmes. Their memories live in the hearts of those who knew them.

Of course, thanks to all of the contributors to the anthology, both for your writing and for your flexibility, commitment, and good cheer in the long, strange trip of bringing this book to fruition. For extra help with literary practicalities, kudos to Kristin Berger, Clare Melley Smith, and especially Tara Masih.

As an independent operator, I am grateful to the Newton Free Library and to the Boston Public Library for congenial surroundings and helpful resources as I worked on this project. Thanks also to the collegial networks of the Association for the Study of Literature and Environment, the Conservation Psychology e-list and community, the American Society for Environmental History, the North American Association for Environmental Education, and the American Academy of Religion. A special bow to NEW-CUE and Barbara Ward Klein, who for many years created a unique and invigorating gathering-space at the intersection of the academic and literary realms. Thanks to Mark Bailey, Kirsten Johanna Allen, and Anne Terashima of Torrey House Press for their work and commitment in helping to bring this book to the world.

And as always, thanks to the music-makers in the world and in my life; and to Carlene, for sharing the music with me.

*Steven Pavlos Holmes*
*Jamaica Plain, Massachusetts*
*April 2013*

## About Torrey House Press

*The economy is a wholly owned subsidiary of the environment, not the other way around.*
—Senator Gaylord Nelson, founder of Earth Day

Headquartered in Salt Lake City and Torrey, Utah, Torrey House Press is an independent book publisher of fiction and nonfiction about the environment, people, cultures, and resource management issues relating to America's wild places. Torrey House Press endeavors to increase appreciation for the importance of natural landscape through the power of pen and story. Through the *2% to the West* program, Torrey House Press donates two percent of sales to not-for-profit environmental organizations and funds a scholarship for up-and-coming writers at colleges throughout the West.

Torrey House Press
**www.torreyhouse.com**
Visit our website for thought-provoking discussion guides, author interviews, and more.

## Forthcoming from Torrey House Press

*Monument Road*
**By Charlie Quimby**
**November 2013 | 978-1-937226-25-1**
After spending a year unwinding his ranch and fending off *the darkening*, Leonard Self knows where he's going to end his life. But the road is winding and he has company.

*My So-Called Ruined Life*
**by Melanie Bishop**
**January 2014 | 978-1-937226-21-3**
With her father on trial for her mother's murder, sixteen-year-old Tate McCoy is determined to prove her life is not ruined. This is the first book in the Tate McCoy Series.

*Wild Rides & Wildflowers: Philosophy and Botany with Bikes*
**by Scott Abbott and Sam Rushforth**
**March 2014 | 978-1-937226-23-7**
Two university professors bike the Great Western Trail, offering often-humorous, sometimes poignant insights into the male psyche, botany, philosophy, and true friendship.

## Available now from Torrey House Press

*A Bushel's Worth: An Ecobiography*
**by Kayann Short**
**August 2013 | 978-1-937226-19-0**
Rooted where the Rocky Mountains meet the prairie, Short's love story of land celebrates our connection to soil and each other, and one community's commitment to keeping a farm a farm.

*Evolved: Chronicles of a Pleistocene Mind*
**by Maximilian Werner**
**June 2013 | 978-1-937226-17-6**
With startling insights, Werner explores how our Pleistocene instincts inform our everyday decisions and behaviors in this modern day Walden.

*The Legend's Daughter*
**by David Kranes**
**May 2013 | 978-1-937226-15-2**
These fast-paced stories set in contemporary Idaho explore intricate dynamics between fathers and sons, unlikely friends, people and place.

*Grind*
**by Mark Maynard**
**December 2012 | 978-1-937226-03-9**
The gritty realism of Hemingway joins the irreverence of Edward Abbey in these linked short stories set in and around Reno, Nevada.

*The Ordinary Truth*
**by Jana Richman**
**November 2012 | 978-1-937226-06-0**
Today's western water wars and one family's secrets divide three generations of women as urban and rural values collide in this contemporary novel.

*Recapture*
**by Erica Olsen**
**October 2012 | 978-1-937226-05-3**
This captivating short story collection explores the canyons, gulches, and vast plains of memory along with the colorful landscapes of the West.

*Tribútary*
**by Barbara K. Richardson**
**September 2012 | 978-1-937226-04-6**
A courageous young woman flees polygamy in 1860s Utah, but finds herself drawn back to the landscapes that shaped her.

*The Plume Hunter*
**by Renée Thompson**
**December 2011 | 978-1-937226-01-5**
Love and lives are lost amid conflict over killing wild birds for women's hats in Oregon and California in the late nineteenth century in this historical novel.

*The Scholar of Moab*
**by Steven L. Peck**
**November 2011 | 978-1937226-02-2**
Philosophy meets satire, poetry, cosmology, and absurdity in this tragicomic brew of magical realism and rural Mormon Utah.

*Crooked Creek*
**by Maximilian Werner**
**June 2011 | 978-1-937226-00-8**
*Blood Meridian* finds *A Farewell to Arms* in this short and beautiful novel set in 1890s Utah.